GENOME

GENOME

The autobiography of a species in 23 chapters

....

Matt Ridley

FOURTH ESTATE • *London*

First published in Great Britain in 1999 by
Fourth Estate Limited
6 Salem Road
London W2 4BU

Copyright © Matt Ridley 1999

The author is grateful to the following publishers for permission to reprint brief extracts: Harvard University Press for four extracts from Nancy Wexler's article in *The code of codes*, edited by D. Kevles and R. Hood (pp. 62–9); Aurum Press for an extract from *The gene hunters* by William Cookson (p. 78); Macmillan Press for extracts from *Philosophical essays* by A. J. Ayer (p. 338) and *What remains to be discovered* by J. Maddox (p. 194); W. H. Freeman for extracts from *Narrow roads of gene land* by W. D. Hamilton (p. 131); Oxford University Press for extracts from *The selfish gene* by Richard Dawkins (p. 122) and *Fatal protein* by Rosalind Ridley and Harry Baker (p. 285); Weidenfeld and Nicolson for an extract from *One renegade cell* by Robert Weinberg (p. 237). The author has made every effort to obtain permission for all other extracts from published work reprinted in this book.

1 3 5 7 9 10 8 6 4 2

A catalogue record for this book is available from the British Library.

ISBN 1-85702-834-1

Typeset by Rowland Phototypesetting Ltd,
Bury St Edmunds, Suffolk

Printed in Great Britain by
Clays Ltd, St Ives plc

For my parents and my children

CONTENTS

ACKNOWLEDGEMENTS

In writing this book, I have disturbed, interrupted, interrogated, emailed and corresponded with a great variety of people, yet I have never once met anything but patience and politeness. I cannot thank everybody by name, but I would like to record my great debts of gratitude to the following: Bill Amos, Rosalind Arden, Christopher Badcock, Rosa Beddington, David Bentley, Ray Blanchard, Sam Brittan, John Burn, Francis Crick, Gerhard Cristofori, Paul Davies, Barry Dickson, Richard Durbin, Jim Edwardson, Myrna Gopnik, Anthony Gottlieb, Dean Hamer, Nick Hastie, Brett Holland, Tony Ingram, Mary James, Harmke Kamminga, Terence Kealey, Arnold Levine, Colin Merritt, Geoffrey Miller, Graeme Mitchison, Anders Moller, Oliver Morton, Kim Nasmyth, Sasha Norris, Mark Pagel, Rose Paterson, David Penny, Marion Petrie, Steven Pinker, Robert Plomin, Anthony Poole, Christine Rees, Janet Rossant, Mark Ridley, Robert Sapolsky, Tom Shakespeare, Ancino Silva, Lee Silver, Tom Strachan, John Sulston, Tim Tully, Thomas Vogt, Jim Watson, Eric Weischaus and Ian Wilmut.

Special thanks to all my colleagues at the International Centre for Life, where we have been trying to bring the genome to life. Without the day-to-day interest and support from them in matters biological and genetic, I doubt I could have written this book. They are Alastair Balls, John Burn, Linda Conlon, Ian Fells, Irene Nyguist, Neil Sullivan, Elspeth Wills and many others.

Parts of two chapters first appeared in newspaper columns and magazine articles. I am grateful to Charles Moore of the *Daily Telegraph* and David Goodhart of *Prospect* for publishing them.

My agent, Felicity Bryan, has been enthusiasm personified throughout. Three editors had more faith in this book when it was just a proposal than (I now admit) I did: Christopher Potter, Marion Manneker and Maarten Carbo.

But to one person I give deeper and more heartfelt gratitude than to all the rest put together: my wife, Anya Hurlbert.

PREFACE

The human genome – the complete set of human genes – comes packaged in twenty-three separate pairs of chromosomes. Of these, twenty-two pairs are numbered in approximate order of size, from the largest (number 1) to the smallest (number 22), while the remaining pair consists of the sex chromosomes: two large X chromosomes in women, one X and one small Y in men. In size, the X comes between chromosomes 7 and 8, whereas the Y is the smallest.

The number 23 is of no significance. Many species, including our closest relatives among the apes, have more chromosomes, and many have fewer. Nor do genes of similar function and type necessarily cluster on the same chromosome. So a few years ago, leaning over a lap-top computer talking to David Haig, an evolutionary biologist, I was slightly startled to hear him say that chromosome 15 was his favourite chromosome. It has all sorts of mischievous genes on it, he explained. I had never thought of chromosomes as having personalities before. They are, after all, merely arbitrary collections of genes. But Haig's chance remark planted an idea in my head and I could not get it out. Why not try to tell the unfolding story of the human genome, now being discovered in detail for the first time, chromosome by chromosome, by picking a gene from each chromosome to fit the story as it is told? Primo Levi did something similar with the periodic table of the elements in his autobiographical short stories. He related each chapter of his life to an element, one that he had had some contact with during the period he was describing.

I began to think about the human genome as a sort of autobiography in its own right – a record, written in 'genetish', of all the vicissitudes and inventions that had characterised the history of our species and its ancestors since the very dawn of life. There are genes that have not changed much since the very first single-celled creatures populated the primeval ooze. There are genes that were developed when our ancestors were worm-like. There are genes that must have first appeared when our ancestors were fish. There are genes that exist in their present form only because of recent epidemics of disease. And there are genes that can be used to write the history of human migrations in the last few thousand years. From four billion years ago to just a few hundred years ago, the genome has been a sort of autobiography for our species, recording the important events as they occurred.

I wrote down a list of the twenty-three chromosomes and next to each I began to list themes of human nature. Gradually and painstakingly I began to find genes that were emblematic of my story. There were frequent frustrations when I could not find a suitable gene, or when I found the ideal gene and it was on the wrong chromosome. There was the puzzle of what to do with the X and Y chromosomes, which I have placed after chromosome 7, as befits the X chromosome's size. You now know why the last chapter of a book that boasts in its subtitle that it has twenty-three chapters is called Chapter 22.

It is, at first glance, a most misleading thing that I have done. I may seem to be implying that chromosome 1 came first, which it did not. I may seem to imply that chromosome 11 is exclusively concerned with human personality, which it is not. There are probably 60,000–80,000 genes in the human genome and I could not tell you about all of them, partly because fewer than 8,000 have been found (though the number is growing by several hundred a month) and partly because the great majority of them are tedious biochemical middle managers.

But what I can give you is a coherent glimpse of the whole: a whistle-stop tour of some of the more interesting sites in the genome and what they tell us about ourselves. For we, this lucky generation, will be the first to read the book that is the genome. Being able to read the genome will tell us more about our origins, our evolution, our nature and our minds than all the efforts of science to date. It will revolutionise anthropology, psychology, medicine, palaeontology and virtually every other science. This is not to claim that everything is in the genes, or that genes matter more than other factors. Clearly, they do not. But they matter, that is for sure.

This is not a book about the Human Genome Project – about mapping and sequencing techniques – but a book about what that project has found. Some time in the year 2000, we shall probably have a rough first draft of the complete human genome. In just a few short years we will have moved from knowing almost nothing about our genes to knowing everything. I genuinely believe that we are living through the greatest intellectual moment in history. Bar none. Some may protest that the human being is more than his genes. I do not deny it. There is much, much more to each of us than a genetic code. But until now human genes were an almost complete mystery. We will be the first generation to penetrate that mystery. We stand on the brink of great new answers but, even more, of great new questions. This is what I have tried to convey in this book.

The second part of this preface is intended as a brief primer, a sort of narrative glossary, on the subject of genes and how they work. I hope that readers will glance through it at the outset and return to it at intervals if they come across technical terms that are not explained. Modern genetics is a formidable thicket of jargon. I have tried hard to use the bare minimum of technical terms in this book, but some are unavoidable.

The human body contains approximately 100 trillion (million million) CELLS, most of which are less than a tenth of a millimetre across. Inside each cell there is a black blob called a NUCLEUS. Inside the nucleus are two complete sets of the human GENOME (except in egg cells and sperm cells, which have one copy each, and red blood cells, which have none). One set of the genome came from the mother and one from the father. In principle, each set includes the same 60,000–80,000 GENES on the same twenty-three CHROMOSOMES. In practice, there are often small and subtle differences between the paternal and maternal versions of each gene, differences that account for blue eyes or brown, for example. When we breed, we pass on one complete set, but only after swapping bits of the paternal and maternal chromosomes in a procedure known as RECOMBINATION.

Imagine that the genome is a book.

There are twenty-three chapters, called CHROMOSOMES.
Each chapter contains several thousand stories, called GENES.
Each story is made up of paragraphs, called EXONS, which are interrupted
    by advertisements called INTRONS.
Each paragraph is made up of words, called CODONS.
Each word is written in letters called BASES.

There are one billion words in the book, which makes it longer than 5,000 volumes the size of this one, or as long as 800 Bibles. If I read the genome out to you at the rate of one word per second for eight hours a day, it would take me a century. If I wrote out the human genome, one letter per centimetre, my text would be as long as the Danube. This is a gigantic document, an immense book, a recipe of extravagant length, and it all fits inside the microscopic nucleus of a tiny cell that fits easily upon the head of a pin.

The idea of the genome as a book is not, strictly speaking, even a metaphor. It is literally true. A book is a piece of digital information, written in linear, one-dimensional and one-directional form and defined by a code that transliterates a small alphabet of signs into a large lexicon of meanings through the order of their groupings. So is a genome. The only complication is that all English books read from left to right, whereas some parts of the genome read from left to right, and some from right to left, though never both at the same time.

(Incidentally, you will not find the tired word 'blueprint' in this book, after this paragraph, for three reasons. First, only architects and engineers use blueprints and even they are giving them up in the computer age, whereas we all use books. Second, blueprints are very bad analogies for genes. Blueprints are two-dimensional maps, not one-dimensional digital codes. Third, blueprints are too literal for genetics, because each part of a blueprint makes an equivalent part of the machine or building; each sentence of a recipe book does not make a different mouthful of cake.)

Whereas English books are written in words of variable length using twenty-six letters, genomes are written entirely in three-letter words, using only four letters: A, C, G and T (which stand for adenine, cytosine, guanine and thymine). And instead of being written on flat pages, they are written on long chains of sugar and phosphate called DNA molecules to which the bases are attached as side rungs. Each chromosome is one pair of (very) long DNA molecules. Laid end to end and stretched out straight, all the chromosomes in a single cell would cover six feet. All the chromosomes in all the cells in one body would cover one hundred billion miles or about two light-days (light travels forty billion miles a day). There are six hundred billion billion miles of human DNA on earth, or enough to stretch from here to the next galaxy.

The genome is a very clever book, because in the right conditions it can both photocopy itself and read itself. The photocopying is known as REPLICATION, and the reading as TRANSLATION. Replication works because of an ingenious property of the four bases: A likes to pair with T, and G with C. So a single strand of DNA can copy itself by assembling a complementary strand with Ts opposite all the As, As opposite all the Ts, Cs opposite all the Gs and Gs opposite all the Cs. In fact, the usual state of DNA is the famous

DOUBLE HELIX of the original strand and its complementary pair intertwined.

To make a copy of the complementary strand therefore brings back the original text. So the sequence ACGT become TGCA in the copy, which transcribes back to ACGT in the copy of the copy. This enables DNA to replicate indefinitely, yet still contain the same information.

Translation is a little more complicated. First the text of a gene is TRANSCRIBED into a copy by the same base-pairing process, but this time the copy is made not of DNA but of RNA, a very slightly different chemical. RNA, too, can carry a linear code and it uses the same letters as DNA except that it uses U, for uracil, in place of T. This RNA copy, called the MESSENGER RNA, is then edited by the excision of all introns and the splicing together of all exons (see above).

The messenger is then befriended by a microscopic machine called a RIBOSOME, itself made partly of RNA. The ribosome moves along the messenger, translating each three-letter codon in turn into one letter of a different alphabet, an alphabet of twenty different AMINO ACIDS, each brought by a different version of a molecule called TRANSFER RNA. Each amino acid is attached to the last to form a chain in the same order as the codons. When the whole message has been translated, the chain of amino acids folds itself up into a distinctive shape that depends on its sequence. It is now known as a PROTEIN.

Almost everything in the body, from hair to hormones, is either made of proteins or made by them. Every protein is a translated gene. In particular, the body's chemical reactions are catalysed by proteins known as ENZYMES. Even the processing, photocopying error-correction and assembly of DNA and RNA molecules themselves – the replication and translation – are done with the help of proteins. Proteins are also responsible for switching genes on and off, by physically attaching themselves to PROMOTER and ENHANCER sequences near the start of a gene's text. Different genes are switched on in different parts of the body.

When genes are replicated, mistakes are sometimes made. A letter (base) is occasionally missed out or the wrong letter inserted. Whole sentences or paragraphs are sometimes duplicated, omitted or reversed. This is known as MUTATION. Many mutations are neither harmful nor beneficial, for instance if they change one codon to another that has the same amino acid 'meaning': there are sixty-four different codons and only twenty amino acids, so many DNA 'words' share the same meaning. Human beings accumulate about one hundred mutations per generation, which may not seem much given that there are more than a million codons in the human genome, but in the wrong place even a single one can be fatal.

All rules have exceptions (including this one). Not all human genes are found on the twenty-three principal chromosomes; a few live inside little blobs called mitochondria and have probably done so ever since mitochondria were free-living bacteria. Not all genes are made of DNA: some viruses use RNA instead. Not all genes are recipes for proteins. Some genes are transcribed into RNA but not translated into protein; the RNA goes directly to work instead either as part of a ribosome or as a transfer RNA. Not all reactions are catalysed by proteins; a few are catalysed by RNA instead. Not every protein comes from a single gene; some are put together from several recipes. Not all of the sixty-four three-letter codons specifies an amino acid: three signify STOP commands instead. And finally, not all DNA spells out genes. Most of it is a jumble of repetitive or random sequences that is rarely or never transcribed: the so-called junk DNA.

That is all you need to know. The tour of the human genome can begin.

CHROMOSOME 1

Life

All forms that perish other forms supply,
(By turns we catch the vital breath and die)
Like bubbles on the sea of matter borne,
They rise, they break, and to that sea return.
Alexander Pope, An Essay on Man

In the beginning was the word. The word proselytised the sea with its message, copying itself unceasingly and forever. The word discovered how to rearrange chemicals so as to capture little eddies in the stream of entropy and make them live. The word transformed the land surface of the planet from a dusty hell to a verdant paradise. The word eventually blossomed and became sufficiently ingenious to build a porridgy contraption called a human brain that could discover and be aware of the word itself.

My porridgy contraption boggles every time I think this thought. In four thousand million years of earth history, I am lucky enough to be alive today. In five million species, I was fortunate enough to be born a conscious human being. Among six thousand million people on the planet, I was privileged enough to be born in the country where the word was discovered. In all of the earth's history, biology and geography, I was born just five years after the moment when, and just two hundred miles from the place where, two members of my own species discovered the structure of DNA and hence uncovered the greatest, simplest and most surprising secret in the universe. Mock my zeal if you wish; consider me a ridiculous materialist for investing such enthusiasm in an acronym. But follow me on a journey back to the very origin of life, and I hope I can convince you of the immense fascination of the word.

'As the earth and ocean were probably peopled with vegetable productions long before the existence of animals; and many families of these animals long before other families of them, shall we conjecture that one and the same kind of living filaments is and has been the cause of all organic life?' asked the polymathic poet and physician Erasmus Darwin in 1794.[1] It was a startling guess for the time, not only in its bold conjecture that all organic life shared the same origin, sixty-five years before his grandson Charles' book on the topic, but for its weird use of the word 'filaments'. The secret of life is indeed a thread.

Yet how can a filament make something live? Life is a slippery thing to define, but it consists of two very different skills: the ability to replicate, and the ability to create order. Living things produce approximate copies of themselves: rabbits produce rabbits, dandelions make dandelions. But rabbits do more than that. They eat grass, transform it into rabbit flesh and somehow build bodies of order and complexity from the random chaos of the world. They do not defy the second law of thermodynamics, which says that in a closed system everything tends from order towards disorder, because rabbits are not closed systems. Rabbits build packets of order and complexity called bodies but at the cost of expending large amounts of energy. In Erwin Schrödinger's phrase, living creatures 'drink orderliness' from the environment.

The key to both of these features of life is information. The ability to replicate is made possible by the existence of a recipe, the information that is needed to create a new body. A rabbit's egg carries the instructions for assembling a new rabbit. But the ability to create order through metabolism also depends on information – the instructions for building and maintaining the equipment that creates the order. An adult rabbit, with its ability to both reproduce and metabolise, is prefigured and presupposed in its living filaments in the same way that a cake is prefigured and presupposed in its recipe. This is an idea that goes right back to Aristotle, who said that the 'concept' of a chicken is implicit in an egg, or that an acorn was literally 'informed' by the plan of an oak tree. When Aristotle's dim perception of information theory, buried under generations of chemistry and physics, re-emerged amid the discoveries of modern genetics, Max Delbruck joked that the Greek sage should be given a posthumous Nobel prize for the discovery of DNA.[2]

The filament of DNA is information, a message written in a code of chemicals, one chemical for each letter. It is almost too good to be true, but the code turns out to be written in a way that we can understand. Just like written English, the genetic code is a linear language, written in a straight line. Just like written English, it is digital, in that every letter bears the same importance. Moreover, the language of DNA is considerably simpler than English, since it has an alphabet of only four letters, conventionally known as A, C, G and T.

Now that we know that genes are coded recipes, it is hard to recall how few people even guessed such a possibility. For the first half of the twentieth century, one question reverberated unanswered through biology: what is a gene? It seemed almost impossibly mysterious. Go back not to 1953, the year of the discovery of DNA's symmetrical structure, but ten years further, to 1943. Those who will do most to crack the mystery, a whole decade later, are working on other things in 1943. Francis Crick is working on the design of naval mines near Portsmouth. At the same time James Watson is just enrolling as an undergraduate at the precocious age of fifteen at the University of Chicago; he is determined to devote his life to ornithology. Maurice Wilkins is helping to design the atom bomb in the United States. Rosalind Franklin is studying the structure of coal for the British government.

In Auschwitz in 1943, Josef Mengele is torturing twins to death in a grotesque parody of scientific inquiry. Mengele is trying to understand heredity, but his eugenics proves not to be the path to enlightenment. Mengele's results will be useless to future scientists.

In Dublin in 1943, a refugee from Mengele and his ilk, the great physicist Erwin Schrödinger is embarking on a series of lectures at Trinity College entitled 'What is life?' He is trying to define a problem. He knows that chromosomes contain the secret of life, but he cannot understand how: 'It is these chromosomes . . . that contain in some kind of code-script the entire pattern of the individual's future development and of its functioning in the mature state.' The gene, he says, is too small to be anything other than a large molecule, an insight that will inspire a generation of scientists, including Crick, Watson, Wilkins and Franklin, to tackle what suddenly seems like a tractable problem. Having thus come tantalisingly close to the answer, though, Schrödinger veers off track. He thinks that the secret of this molecule's ability to carry heredity lies in his beloved quantum theory, and is pursuing that obsession down what will prove to be a blind alley. The secret of life has nothing to do with quantum states. The answer will not come from physics.[3]

In New York in 1943, a sixty-six-year-old Canadian scientist, Oswald Avery, is putting the finishing touches to an experiment that will decisively identify DNA as the chemical manifestation of heredity. He has proved in a series of ingenious experiments that a pneumonia bacterium can be transformed from a harmless to a virulent strain merely by absorbing a simple chemical solution. By 1943, Avery has concluded that the transforming substance, once purified, is DNA. But he will couch his conclusions in such cautious language for publication that few will take notice until much later. In a letter to his brother Roy written in May 1943, Avery is only slightly less cautious:[4]

If we are right, and of course that's not yet proven, then it means that nucleic acids [DNA] are not merely structurally important but functionally active substances in determining the biochemical activities and specific characteristics of cells – and that by means of a known chemical substance it is possible to induce predictable and hereditary changes in cells. That is something that has long been the dream of geneticists.

Avery is almost there, but he is still thinking along chemical lines. 'All life is chemistry', said Jan Baptist van Helmont in 1648, guessing. At least some life is chemistry, said Friedrich Wöhler in 1828 after synthesising urea from ammonium chloride and silver cyanide, thus breaking the hitherto sacrosanct divide between the chemical and biological worlds: urea was something that only living things had produced before. That life is chemistry is true but boring, like saying that football is physics. Life, to a rough approximation, consists of the chemistry of three atoms, hydrogen, carbon and oxygen, which between them make up ninety-eight per cent of all atoms in living beings. But it is the emergent properties of life – such as heritability – not the constituent parts that are interesting. Avery cannot conceive what it is about DNA that enables it to hold the secret of heritable properties. The answer will not come from chemistry.

In Bletchley, in Britain, in 1943, in total secrecy, a brilliant mathematician, Alan Turing, is seeing his most incisive insight turned into physical reality. Turing has argued that numbers can compute numbers. To crack the Lorentz encoding machines of the German forces, a computer called Colossus has been built based on Turing's principles: it is a universal machine with a modifiable stored program. Nobody realises it at the time, least of all Turing, but he is probably closer to the mystery of life than anybody else. Heredity is a modifiable stored program; metabolism is a universal machine. The recipe that links them is a code, an abstract message that can be embodied in a chemical, physical or even immaterial form. Its secret is that it can cause itself to be replicated. Anything that can use the resources of the world to get copies of itself made is alive; the most likely form for such a thing to take is a digital message – a number, a script or a word.[5]

In New Jersey in 1943, a quiet, reclusive scholar named Claude Shannon is ruminating about an idea he had first had at Princeton a few years earlier. Shannon's idea is that information and entropy are opposite faces of the same coin and that both have an intimate link with energy. The less entropy a system has, the more information it contains. A steam engine parcels out entropy to generate energy because of the information injected into it by its designer. So does a human body. Aristotle's information theory meets Newton's physics in Shannon's brain. Like Turing, Shannon has no thoughts about biology. But his insight is of more relevance to the question of what is life than a mountain of chemistry and physics. Life, too, is digital information written in DNA.[6]

In the beginning was the word. The word was not DNA. That came afterwards, when life was already established, and when it had divided the labour between two separate activities: chemical work and information storage, metabolism and replication. But DNA contains a record of the word, faithfully transmitted through all subsequent aeons to the astonishing present.

Imagine the nucleus of a human egg beneath the microscope. Arrange the twenty-three chromosomes, if you can, in order of size, the biggest on the left and the smallest on the right. Now zoom in on the largest chromosome, the one called, for purely arbitrary reasons, chromosome 1. Every chromosome has a long arm and a short arm separated by a pinch point known as a centromere. On the long arm of chromosome 1, close to the centromere, you will find, if you read it carefully, that there is a sequence of 120 letters – As, Cs, Gs and Ts – that repeats over and over again. Between each repeat there lies a stretch of more random text, but the 120-letter paragraph keeps coming back like a familiar theme tune, in all more than 100 times. This short paragraph is perhaps as close as we can get to an echo of the original word.

This 'paragraph' is a small gene, probably the single most active gene in the human body. Its 120 letters are constantly being copied into a short filament of RNA. The copy is known as 5S RNA. It sets up residence with a lump of proteins and other RNAs, carefully intertwined, in a ribosome, a machine whose job is to translate DNA recipes into proteins. And it is proteins that enable DNA to replicate. To paraphrase Samuel Butler, a protein is just a gene's way of making another gene; and a gene is just a protein's way of making another protein. Cooks need recipes, but recipes also need cooks. Life consists of the interplay of two kinds of chemicals: proteins and DNA.

Protein represents chemistry, living, breathing, metabolism and behaviour – what biologists call the phenotype. DNA represents information, replication, breeding, sex – what biologists call the genotype. Neither can exist without the other. It is the classic case of chicken and egg: which came first, DNA or protein? It cannot have been DNA, because DNA is a helpless, passive piece of mathematics, which catalyses no chemical reactions. It cannot have been protein, because protein is pure chemistry with no known way of copying itself accurately. It seems impossible either that DNA invented protein or vice versa. This might have remained a baffling and strange conundrum had not the word left a trace of itself faintly drawn on the filament of life. Just as we now know that eggs came long before chickens (the reptilian ancestors of all birds laid eggs), so there is growing evidence that RNA came before proteins.

RNA is a chemical substance that links the two worlds of DNA and protein. It is used mainly in the translation of the message from the alphabet of DNA to the alphabet of proteins. But in the way it behaves, it leaves little doubt that it is the ancestor of both. RNA was Greece to DNA's Rome: Homer to her Virgil.

RNA was the word. RNA left behind five little clues to its priority over both protein and DNA. Even today, the ingredients of DNA are made by modifying the ingredients of RNA, not by a more direct route. Also DNA's letter Ts are made from RNA's letter Us. Many modern enzymes, though made of protein, rely on small molecules of RNA to make them work. Moreover, RNA, unlike DNA and protein, can copy itself without assistance: give it the right ingredients and it will stitch them together into a message. Wherever you look in the cell, the most primitive and basic functions require the presence of RNA. It is an RNA-dependent enzyme that takes the message, made of RNA, from the gene. It is an

RNA-containing machine, the ribosome, that translates that message, and it is a little RNA molecule that fetches and carries the
amino acids for the translation of the gene's message. But above
all, RNA – unlike DNA – can act as a catalyst, breaking up and
joining other molecules including RNAs themselves. It can cut
them up, join the ends together, make some of its own building
blocks, and elongate a chain of RNA. It can even operate on itself,
cutting out a chunk of text and splicing the free ends together
again.[7]

The discovery of these remarkable properties of RNA in the
early 1980s, made by Thomas Cech and Sidney Altman, transformed
our understanding of the origin of life. It now seems probable that
the very first gene, the 'ur-gene', was a combined replicator–catalyst,
a word that consumed the chemicals around it to duplicate itself. It
may well have been made of RNA. By repeatedly selecting random
RNA molecules in the test tube based on their ability to catalyse
reactions, it is possible to 'evolve' catalytic RNAs from scratch –
almost to rerun the origin of life. And one of the most surprising
results is that these synthetic RNAs often end up with a stretch of
RNA text that reads remarkably like part of the text of a ribosomal
RNA gene such as the 5S gene on chromosome 1.

Back before the first dinosaurs, before the first fishes, before the
first worms, before the first plants, before the first fungi, before the
first bacteria, there was an RNA world – probably somewhere
around four billion years ago, soon after the beginning of planet
earth's very existence and when the universe itself was only ten
billion years old. We do not know what these 'ribo-organisms'
looked like. We can only guess at what they did for a living, chemi-
cally speaking. We do not know what came before them. We can
be pretty sure that they once existed, because of the clues to RNA's
role that survive in living organisms today.[8]

These ribo-organisms had a big problem. RNA is an unstable
substance, which falls apart within hours. Had these organisms ven-
tured anywhere hot, or tried to grow too large, they would have
faced what geneticists call an error catastrophe – a rapid decay of

the message in their genes. One of them invented by trial and error a new and tougher version of RNA called DNA and a system for making RNA copies from it, including a machine we'll call the proto-ribosome. It had to work fast and it had to be accurate. So it stitched together genetic copies three letters at a time, the better to be fast and accurate. Each threesome came flagged with a tag to make it easier for the proto-ribosome to find, a tag that was made of amino acid. Much later, those tags themselves became joined together to make proteins and the three-letter word became a form of code for the proteins – the genetic code itself. (Hence to this day, the genetic code consists of three-letter words, each spelling out a particular one of twenty amino acids as part of a recipe for a protein.) And so was born a more sophisticated creature that stored its genetic recipe on DNA, made its working machines of protein and used RNA to bridge the gap between them.

Her name was Luca, the Last Universal Common Ancestor. What did she look like, and where did she live? The conventional answer is that she looked like a bacterium and she lived in a warm pond, possibly by a hot spring, or in a marine lagoon. In the last few years it has been fashionable to give her a more sinister address, since it became clear that the rocks beneath the land and sea are impregnated with billions of chemical-fuelled bacteria. Luca is now usually placed deep underground, in a fissure in hot igneous rocks, where she fed on sulphur, iron, hydrogen and carbon. To this day, the surface life on earth is but a veneer. Perhaps ten times as much organic carbon as exists in the whole biosphere is in thermophilic bacteria deep beneath the surface, where they are possibly responsible for generating what we call natural gas.[9]

There is, however, a conceptual difficulty about trying to identify the earliest forms of life. These days it is impossible for most creatures to acquire genes except from their parents, but that may not always have been so. Even today, bacteria can acquire genes from other bacteria merely by ingesting them. There might once have been widespread trade, even burglary, of genes. In the deep past chromosomes were probably numerous and short, containing just one gene each, which could be lost or gained quite easily. If this was so, Carl Woese points out, the organism was not yet an enduring entity. It was a temporary team of genes. The genes that ended up in all of us may therefore have come from lots of different 'species' of creature and it is futile to try to sort them into different lineages. We are descended not from one ancestral Luca, but from the whole community of genetic organisms. Life, says Woese, has a physical history, but not a genealogical one.[10]

You can look on such a conclusion as a fuzzy piece of comforting, holistic, communitarian philosophy — we are all descended from society, not from an individual species — or you can see it as the ultimate proof of the theory of the selfish gene: in those days, even more than today, the war was carried on between genes, using organisms as temporary chariots and forming only transient alliances; today it is more of a team game. Take your pick.

Even if there were lots of Lucas, we can still speculate about where they lived and what they did for a living. This is where the second problem with the thermophilic bacteria arises. Thanks to some brilliant detective work by three New Zealanders published in 1998, we can suddenly glimpse the possibility that the tree of life, as it appears in virtually every textbook, may be upside down. Those books assert that the first creatures were like bacteria, simple cells with single copies of circular chromosomes, and that all other living things came about when teams of bacteria ganged together to make complex cells. It may much more plausibly be the exact reverse. The very first modern organisms were not like bacteria; they did not live in hot springs or deep-sea volcanic vents. They were much more like protozoa: with genomes fragmented into several linear chromosomes rather than one circular one, and 'polyploid' — that is, with several spare copies of every gene to help with the correction of spelling errors. Moreover, they would have liked cool climates. As Patrick Forterre has long argued, it now looks as if bacteria came later, highly specialised and simplified descendants of the Lucas, long after the invention of the DNA-protein world. Their trick was to drop much of the equipment of the RNA world specifically to enable them to live in hot places. It is we that have retained the primitive molecular features of the Lucas in our cells; bacteria are much more 'highly evolved' than we are.

This strange tale is supported by the existence of molecular 'fossils' – little bits of RNA that hang about inside the nucleus of your cells doing unnecessary things such as splicing themselves out of genes: guide RNA, vault RNA, small nuclear RNA, small nucleolar RNA, self-splicing introns. Bacteria have none of these, and it is more parsimonious to believe that they dropped them rather than we invented them. (Science, perhaps surprisingly, is supposed to treat simple explanations as more probable than complex ones unless given reason to think otherwise; the principle is known in logic as Occam's razor.) Bacteria dropped the old RNAs when they invaded hot places like hot springs or subterranean rocks where temperatures can reach 170 °C – to minimise mistakes caused by heat, it paid to simplify the machinery. Having dropped the RNAs, bacteria found their new streamlined cellular machinery made them good at competing in niches where speed of reproduction was an advantage – such as parasitic and scavenging niches. We retained those old RNAs, relics of machines long superseded, but never entirely thrown away. Unlike the massively competitive world of bacteria, we – that is all animals, plants and fungi – never came under such fierce competition to be quick and simple. We put a premium instead on being complicated, in having as many genes as possible, rather than a streamlined machine for using them.[11]

The three-letter words of the genetic code are the same in every creature. CGA means arginine and GCG means alanine – in bats, in beetles, in beech trees, in bacteria. They even mean the same in the misleadingly named archaebacteria living at boiling temperatures in sulphurous springs thousands of feet beneath the surface of the Atlantic ocean or in those microscopic capsules of deviousness called viruses. Wherever you go in the world, whatever animal, plant, bug or blob you look at, if it is alive, it will use the same dictionary and know the same code. All life is one. The genetic code, bar a few tiny local aberrations, mostly for unexplained reasons in the ciliate protozoa, is the same in every creature. We all use exactly the same language.

This means – and religious people might find this a useful argument – that there was only one creation, one single event when life was born. Of course, that life might have been born on a different planet and seeded here by spacecraft, or there might even have been thousands of kinds of life at first, but only Luca survived in the ruthless free-for-all of the primeval soup. But until the genetic code was cracked in the 1960s, we did not know what we now know: that all life is one; seaweed is your distant cousin and anthrax one of your advanced relatives. The unity of life is an empirical fact. Erasmus Darwin was outrageously close to the mark: 'One and the same kind of living filaments has been the cause of all organic life.'

In this way simple truths can be read from the book that is the genome: the unity of all life, the primacy of RNA, the chemistry of the very earliest life on the planet, the fact that large, single-celled creatures were probably the ancestors of bacteria, not vice versa. We have no fossil record of the way life was four billion years ago. We have only this great book of life, the genome. The genes in the cells of your little finger are the direct descendants of the first replicator molecules; through an unbroken chain of tens of billions of copyings, they come to us today still bearing a digital message that has traces of those earliest struggles of life. If the human genome can tell us things about what happened in the primeval soup, how much more can it tell us about what else happened during the succeeding four million millennia. It is a record of our history written in the code for a working machine.

CHROMOSOME 2

Species

Man with all his noble qualities still bears in his bodily
frame the indelible stamp of his lowly origin.

Charles Darwin

Sometimes the obvious can stare you in the face. Until 1955, it was
agreed that human beings had twenty-four pairs of chromosomes.
It was just one of those facts that everybody knew was right. They
knew it was right because in 1921 a Texan named Theophilus Painter
had sliced thin sections off the testicles of two black men and one
white man castrated for insanity and 'self-abuse', fixed the slices in
chemicals and examined them under the microscope. Painter tried
to count the tangled mass of unpaired chromosomes he could see
in the spermatocytes of the unfortunate men, and arrived at the
figure of twenty-four. 'I feel confident that this is correct,' he said.
Others later repeated his experiment in other ways. All agreed the
number was twenty-four.

For thirty years, nobody disputed this 'fact'. One group of scientists abandoned their experiments on human liver cells because they
could only find twenty-three pairs of chromosomes in each cell.

Another researcher invented a method of separating the chromosomes, but still he thought he saw twenty-four pairs. It was not until 1955, when an Indonesian named Joe-Hin Tjio travelled from Spain to Sweden to work with Albert Levan, that the truth dawned. Tjio and Levan, using better techniques, plainly saw twenty-three pairs. They even went back and counted twenty-three pairs in photographs in books where the caption stated that there were twenty-four pairs. There are none so blind as do not wish to see.[1]

It is actually rather surprising that human beings do not have twenty-four pairs of chromosomes. Chimpanzees have twenty-four pairs of chromosomes; so do gorillas and orang utans. Among the apes we are the exception. Under the microscope, the most striking and obvious difference between ourselves and all the other great apes is that we have one pair less. The reason, it immediately becomes apparent, is not that a pair of ape chromosomes has gone missing in us, but that two ape chromosomes have fused together in us. Chromosome 2, the second biggest of the human chromosomes, is in fact formed from the fusion of two medium-sized ape chromosomes, as can be seen from the pattern of black bands on the respective chromosomes.

Pope John-Paul II, in his message to the Pontifical Academy of Sciences on 22 October 1996, argued that between ancestral apes and modern human beings, there was an 'ontological discontinuity' – a point at which God injected a human soul into an animal lineage. Thus can the Church be reconciled to evolutionary theory. Perhaps the ontological leap came at the moment when two ape chromosomes were fused, and the genes for the soul lie near the middle of chromosome 2.

The pope notwithstanding, the human species is by no means the pinnacle of evolution. Evolution has no pinnacle and there is no such thing as evolutionary progress. Natural selection is simply the process by which life-forms change to suit the myriad opportunities afforded by the physical environment and by other life-forms. The black-smoker bacterium, living in a sulphurous vent on the floor of the Atlantic ocean and descended from a stock of bacteria that parted company with our ancestors soon after Luca's day, is arguably more highly evolved than a bank clerk, at least at the genetic level. Given that it has a shorter generation time, it has had more time to perfect its genes.

This book's obsession with the condition of one species, the human species, says nothing about that species' importance. Human beings are of course unique. They have the most complicated biological machine on the planet perched between their ears. But complexity is not everything, and it is not the goal of evolution. Every species on the planet is unique. Uniqueness is a commodity in oversupply. None the less, I propose to try to probe this human uniqueness in this chapter, to uncover the causes of our idiosyncrasy as a species. Forgive my parochial concerns. The story of a briefly abundant hairless primate originating in Africa is but a footnote in the history of life, but in the history of the hairless primate it is central. What exactly is the unique selling point of our species?

Human beings are an ecological success. They are probably the most abundant large animal on the whole planet. There are nearly six billion of them, amounting collectively to something like 300 million tonnes of biomass. The only large animals that rival or exceed this quantity are ones we have domesticated – cows, chickens and sheep – or that depend on man-made habitats: sparrows and rats. By contrast, there are fewer than a thousand mountain gorillas in the world and even before we started slaughtering them and eroding their habitat there may not have been more than ten times that number. Moreover, the human species has shown a remarkable capacity for colonising different habitats, cold or hot, dry or wet, high or low, marine or desert. Ospreys, barn owls and roseate terns are the only other large species to thrive in every continent except Antarctica and they remain strictly confined to certain habitats. No doubt, this ecological success of the human being comes at a high price and we are doomed to catastrophe soon enough: for a successful species we are remarkably pessimistic about the future. But for now we are a success.

Yet the remarkable truth is that we come from a long line of failures. We are apes, a group that almost went extinct fifteen million years ago in competition with the better-designed monkeys. We are primates, a group of mammals that almost went extinct forty-five million years ago in competition with the better-designed rodents. We are synapsid tetrapods, a group of reptiles that almost went extinct 200 million years ago in competition with the better-designed dinosaurs. We are descended from limbed fishes, which almost went extinct 360 million years ago in competition with the better-designed ray-finned fishes. We are chordates, a phylum that survived the Cambrian era 500 million years ago by the skin of its teeth in competition with the brilliantly successful arthropods. Our ecological success came against humbling odds.

In the four billion years since Luca, the word grew adept at building what Richard Dawkins has called 'survival machines': large, fleshy entities known as bodies that were good at locally reversing entropy the better to replicate the genes within them. They had done this by a venerable and massive process of trial and error, known as natural selection. Trillions of new bodies had been built, tested and enabled to breed only if they met increasingly stringent criteria for survival. At first, this had been a simple business of chemical efficiency: the best bodies were cells that found ways to convert other chemicals into DNA and protein. This phase lasted for about three billion years and it seemed as if life on earth, whatever it might do on other planets, consisted of a battle between competing strains of amoebae. Three billion years during which trillions of trillions of single-celled creatures lived, each one reproducing and dying every few days or so, amounts to a big heap of trial and error.

But it turned out that life was not finished. About a billion years ago, there came, quite suddenly, a new world order, with the invention of bigger, multicellular bodies, a sudden explosion of large creatures. Within the blink of a geological eye (the so-called Cambrian explosion may have lasted a mere ten or twenty million years), there were vast creatures of immense complexity: scuttling trilobites nearly a foot long; slimy worms even longer; waving algae half a yard across. Single-celled creatures still dominated, but these great unwieldy forms of giant survival machines were carving out a niche for themselves. And, strangely, these multicellular bodies had hit upon a sort of accidental progress. Although there were occasional setbacks caused by meteorites crashing into the earth from space, which had an unfortunate tendency to extirpate the larger and more complex forms, there was a trend of sorts discernible. The longer animals existed, the more complex *some* of them became. In particular, the brains of the brainiest animals were bigger and bigger in each successive age: the biggest brains in the Paleozoic were smaller than the biggest in the Mesozoic, which were smaller than the biggest in the Cenozoic, which were smaller than the biggest present now. The genes had found a way to delegate their ambitions, by building bodies capable not just of survival, but of intelligent behaviour as well. Now, if a gene found itself in an animal threatened by winter storms, it could rely on its body to do something clever like migrate south or build itself a shelter.

Our breathless journey from four billion years ago brings us to just ten million years ago. Past the first insects, fishes, dinosaurs and birds to the time when the biggest-brained creature on the planet (corrected for body size) was probably our ancestor, an ape. At that point, ten million years before the present, there probably lived at least two species of ape in Africa, though there may have been more. One was the ancestor of the gorilla, the other the common ancestor of the chimpanzee and the human being. The gorilla's ancestor had probably taken to the montane forests of a string of central African volcanoes, cutting itself off from the genes of other apes. Some time over the next five million years the other species gave rise to two different descendant species in the split that led to human beings and to chimpanzees.

The reason we know this is that the story is written in the genes. As recently as 1950 the great anatomist J. Z. Young could write that it was still not certain whether human beings descended from a common ancestor with apes, or from an entirely different group of primates separated from the ape lineage more than sixty million years ago. Others still thought the orang utan might prove our closest cousin.[2] Yet we now know not only that chimpanzees separated from the human line after gorillas did, but that the chimp–human split occurred not much more than ten, possibly even less than five, million years ago. The rate at which genes randomly accumulate spelling changes gives a firm indication of relationships between species. The spelling differences between gorilla and chimp are greater than the spelling differences between chimp and human being – in every gene, protein sequence or random stretch of DNA that you care to look at. At its most prosaic this means that a hybrid of human and chimpanzee DNA separates into its constituent strands at a higher temperature than do hybrids of chimp and gorilla DNA, or of gorilla and human DNA.

Calibrating the molecular clock to give an actual date in years is much more difficult. Because apes are long-lived and breed at a comparatively advanced age, their molecular clocks tick rather slowly (the spelling mistakes are picked up mostly at the moment of replication, at the creation of an egg or sperm). But it is not clear exactly how much to correct the clock for this factor; nor do all genes agree. Some stretches of DNA seem to imply an ancient split between chimps and human beings; others, such as the mitochondria, suggest a more recent date. The generally accepted range is five to ten million years.[3]

Apart from the fusion of chromosome 2, visible differences between chimp and human chromosomes are few and tiny. In thirteen chromosomes no visible differences of any kind exist. If you select at random any 'paragraph' in the chimp genome and compare it with the comparable 'paragraph' in the human genome, you will find very few 'letters' are different: on average, less than two in every hundred. We are, to a ninety-eight per cent approximation, chimpanzees, and they are, with ninety-eight per cent confidence limits, human beings. If that does not dent your self-esteem, consider that chimpanzees are only ninety-seven per cent gorillas; and humans are also ninety-seven per cent gorillas. In other words we are more chimpanzee-like than gorillas are.

How can this be? The differences between me and a chimp are immense. It is hairier, it has a different shaped head, a different

shaped body, different limbs, makes different noises. There is nothing about chimpanzees that looks ninety-eight per cent like me. Oh
really? Compared with what? If you took two Plasticene models of
a mouse and tried to turn one into a chimpanzee, the other into a
human being, most of the changes you would make would be the
same. If you took two Plasticene amoebae and turned one into a
chimpanzee, the other into a human being, almost all the changes
you would make would be the same. Both would need thirty-two
teeth, five fingers, two eyes, four limbs and a liver. Both would need
hair, dry skin, a spinal column and three little bones in the middle
ear. From the perspective of an amoeba, or for that matter a fertilised
egg, chimps and human beings are ninety-eight per cent the same.
There is no bone in the chimpanzee body that I do not share. There
is no known chemical in the chimpanzee brain that cannot be found
in the human brain. There is no known part of the immune system,
the digestive system, the vascular system, the lymph system or the
nervous system that we have and chimpanzees do not, or vice versa.

There is not even a brain lobe in the chimpanzee brain that we
do not share. In a last, desperate defence of his species against the
theory of descent from the apes, the Victorian anatomist Sir Richard
Owen once claimed that the hippocampus minor was a brain lobe
unique to human brains, so it must be the seat of the soul and the
proof of divine creation. He could not find the hippocampus minor
in the freshly pickled brains of gorillas brought back from the Congo
by the adventurer Paul du Chaillu. Thomas Henry Huxley furiously
responded that the hippocampus minor was there in ape brains.
'No, it wasn't', said Owen. 'Was, too', said Huxley. Briefly, in 1861,
the 'hippocampus question' was all the rage in Victorian London
and found itself satirised in *Punch* and Charles Kingsley's novel *The
water babies*. Huxley's point – of which there are loud modern echoes
– was more than just anatomy:[4] 'It is not I who seek to base Man's
dignity upon his great toe, or insinuate that we are lost if an Ape
has a hippocampus minor. On the contrary, I have done my best
to sweep away this vanity.' Huxley, by the way, was right.

After all, it is less than 300,000 human generations since the common ancestor of both species lived in central Africa. If you held hands with your mother, and she held hands with hers, and she with hers, the line would stretch only from London to Leeds before you were holding hands with the 'missing link' – the common ancestor with chimpanzees. Five million years is a long time, but evolution works not in years but in generations. Bacteria can pack in that many generations in just twenty-five years.

What did the missing link look like? By scratching back through the fossil record of direct human ancestors, scientists are getting remarkably close to knowing. The closest they have come is probably a little ape-man skeleton called Ardipithecus from just over four million years ago. Although a few scientists have speculated that Ardipithecus predates the missing link, it seems unlikely: the creature had a pelvis designed chiefly for upright walking; to modify that back to the gorilla-like pelvis design in the chimpanzee's lineage would have been drastically improbable. We need to find a fossil several million years older to be sure we are looking at a common ancestor of us and chimps. But we can guess, from Ardipithecus, what the missing link looked like: its brain was probably smaller than a modern chimp's. Its body was at least as agile on two legs as a modern chimp's. Its diet, too, was probably like a modern chimp's: mostly fruit and vegetation. Males were considerably bigger than females. It is hard, from the perspective of human beings, not to think of the missing link as more chimp-like than human-like. Chimps might disagree, of course, but none the less it seems as if our lineage has seen grosser changes than theirs.

Like every ape that had ever lived, the missing link was probably a forest creature: a model, modern, Pliocene ape at home among the trees. At some point, its population became split in half. We know this because the separation of two parts of a population is often the event that sparks speciation: the two daughter populations gradually diverge in genetic make-up. Perhaps it was a mountain range, or a river (the Congo river today divides the chimpanzee from its sister species, the bonobo), or the creation of the western Rift Valley itself about five million years ago that caused the split,

leaving human ancestors on the dry, eastern side. The French paleon-
tologist Yves Coppens has called this latter theory 'East Side Story'.
Perhaps, and the theories are getting more far-fetched now, it was
the newly formed Sahara desert that isolated our ancestor in North
Africa, while the chimp's ancestor remained to the south. Perhaps
the sudden flooding, five million years ago, of the then-dry Mediterranean basin by a gigantic marine cataract at Gibraltar, a cataract
one thousand times the volume of Niagara, suddenly isolated a small
population of missing links on some large Mediterranean island,
where they took to a life of wading in the water after fish and
shellfish. This 'aquatic hypothesis' has all sorts of things going for
it except hard evidence.

Whatever the mechanism, we can guess that our ancestors were
a small, isolated band, while those of the chimpanzees were the
main race. We can guess this because we know from the genes that
human beings went through a much tighter genetic bottleneck (i.e.,
a small population size) than chimpanzees ever did: there is much less
random variability in the human genome than the chimp genome.[5]

So let us picture this isolated group of animals on an island, real
or virtual. Becoming inbred, flirting with extinction, exposed to the
forces of the genetic founder effect (by which small populations
can have large genetic changes thanks to chance), this little band of
apes shares a large mutation: two of their chromosomes have
become fused. Henceforth they can breed only with their own kind,
even when the 'island' rejoins the 'mainland'. Hybrids between them
and their mainland cousins are infertile. (I'm guessing again – but
scientists show remarkably little curiosity about the reproductive
isolation of our species: can we breed with chimps or not?)

By now other startling changes have begun to come about. The
shape of the skeleton has changed to allow an upright posture and
a bipedal method of walking, which is well suited to long distances
in even terrain; the knuckle-walking of other apes is better suited
to shorter distances over rougher terrain. The skin has changed, too.
It is becoming less hairy and, unusually for an ape, it sweats profusely
in the heat. These features, together with a mat of hair to shade the

head and a radiator-shunt of veins in the scalp, suggest that our ancestors were no longer in a cloudy and shaded forest; they were walking in the open, in the hot equatorial sun.[6]

Speculate as much as you like about the ecology that selected such a dramatic change in our ancestral skeleton. Few suggestions can be ruled out or in. But by far the most plausible cause of these changes is the isolation of our ancestors in a relatively dry, open grassland environment. The habitat had come to us, not vice versa: in many parts of Africa the savannah replaced the forest about this time. Some time later, about 3.6 million years ago, on freshly wetted volcanic ash recently blown from the Sadiman volcano in what is now Tanzania, three hominids walked purposefully from south to north, the larger one in the lead, the middle-sized one stepping in the leader's footsteps and the small one, striding out to keep up, just a little to the left of the others. After a while, they paused and turned to the west briefly, then walked on, as upright as you or me. The Laetoli fossilised footprints tell as plain a tale of our ancestors' upright walking as we could wish for.

Yet we still know too little. Were the Laetoli ape-people a male, a female and a child or a male and two females? What did they eat? What habitat did they prefer? Eastern Africa was certainly growing drier as the Rift Valley interrupted the circulation of moist winds from the west, but that does not mean they sought dry places. Indeed, our need for water, our tendency to sweat, our peculiar adaptation to a diet rich in the oils and fats of fish and other factors (even our love of beaches and water sports) hint at something of an aquatic preference. We are really rather good at swimming. Were we at first to be found in riverine forests or at the edges of lakes?

In due time, human beings would turn dramatically carnivorous. A whole new species of ape-man, indeed several species, would appear before that, descendants of Laetoli-like creatures, but not ancestors of people, and probably dedicated vegetarians. They are called the robust australopithecines. The genes cannot help us here, because the robusts were dead ends. Just as we would never have known about our close cousinship with chimps if we could not read genes, so we would never have been aware of the existence of our many and closer australopithecine cousins if we had not found fossils (by 'we', I mean principally the Leakey family, Donald Johanson and others). Despite their robust name (which refers only to their heavy jaws), robust australopithecines were little creatures, smaller than chimps and stupider, but erect of posture and heavy of face: equipped with massive jaws supported by giant muscles. They were into chewing – probably grasses and other tough plants. They had lost their canine teeth the better to chew from side to side. Eventually, they became extinct, some time around a million years ago. We may never know much more about them. Perhaps we ate them.

After all, by then our ancestors were bigger animals, as big as modern people, maybe slightly bigger: strapping lads who would grow to nearly six foot, like the famous skeleton of the Nariokotome boy of 1.6 million years ago described by Alan Walker and Richard Leakey.[7] They had begun to use stone tools as substitutes for tough teeth. Perfectly capable of killing and eating a defenceless robust australopithecine – in the animal world, cousins are not safe: lions kill leopards and wolves kill coyotes – these thugs had thick craniums and stone weapons (the two probably go together). Some competitive impulse was now marching the species towards its future explosive success, though nobody directed it – the brain just kept getting bigger and bigger. Some mathematical masochist has calculated that the brain was adding 150 million brain cells every hundred thousand years, the sort of useless statistic beloved of a Soviet tourist guide. Big brains, meat eating, slow development, the 'neotenised' retention into adulthood of childhood characters (bare skin, small jaws and a domed cranium) – all these went together. Without the meat, the protein-hungry brain was an expensive luxury. Without the neotenised skull, there was no cranial space for the brain. Without the slow development, there was no time for learning to maximise the advantages of big brains.

Driving the whole process, perhaps, was sexual selection. Besides the changes to brains, another remarkable change was going on. Females were getting big relative to males. Whereas in modern chimpanzees and australopithecines and the earliest ape-men fossils, males were one-and-a-half times the size of females, in modern people the ratio is much less. The steady decline of that ratio in the fossil record is one of the most overlooked features of our pre-history. What it means is that the mating system of the species was changing. The promiscuity of the chimp, with its short sexual li-aisons, and the harem polygamy of the gorilla, were being replaced with something much more monogamous: a declining ratio of sexual dimorphism is unambiguous evidence for that. But in a more monog-amous system, there would now be pressure on each sex to choose its mate carefully; in polygamy, only the female is choosy. Long pair-bonds shackled each ape-man to its mate for much of its reproductive life: quality rather than quantity was suddenly important. For males it was suddenly vital to choose young mates, because young females had longer reproductive lives ahead of them. A preference for youth-ful, neotenous characters in either sex meant a preference for the large, domed cranium of youth, so it would have begun the drive towards bigger brains and all that followed therefrom.

Pushing us towards habitual monogamy, or at least pulling us further into it, was the sexual division of labour over food. Like no other species on the planet, we had invented a unique partnership between the sexes. By sharing plant food gathered by women, men had won the freedom to indulge the risky luxury of hunting for meat. By sharing hunted meat gathered by men, women had won access to high-protein, digestible food without having to abandon their young in seeking it. It meant that our species had a way of living on the dry plains of Africa that cut the risk of starvation; when meat was scarce, plant food filled the gap; when nuts and fruits were scarce, meat filled the gap. We had therefore acquired a high-protein diet without the intense specialisation for hunting of the big cats.

The habit acquired through the sexual division of labour had spread to other aspects of life. We had become compulsively good at sharing things, which had the new benefit of allowing each indi-vidual to specialise. It was this division of labour among specialists, unique to our species, that was the key to our ecological success, because it allowed the growth of technology. Today we live in societies that express the division of labour in ever more inventive and global ways.[8]

From the here and now, these trends have a certain coherence. Big brains needed meat (vegans today avoid protein-deficiency only by eating pulses); food sharing allowed a meaty diet (because it freed the men to risk failure in pursuit of game); food sharing demanded big brains (without detailed calculating memories, you could be easily cheated by a free rider); the sexual division of labour promoted monogamy (a pair-bond being now an economic unit); monogamy led to neotenous sexual selection (by putting a premium on youthfulness in mates). And so on, round and round the theories we go in a spiral of comforting justification, proving how we came to be as we are. We have built a scientific house of cards on the flimsiest foundations of evidence, but we have reason to believe that it will one day be testable. The fossil record will tell us only a little about behaviour; the bones are too dry and random to speak. But the genetic record will tell us more. Natural selection is the process by which genes change their sequences. In the process of changing, though, those genes laid down a record of our four-billion year biography as a biological lineage. They are, if we only know how to read them, a more valuable source of information on our past than the manuscripts of the Venerable Bede. As I shall argue, a record of our past is etched into our genes.

Some two per cent of the genome tells the story of our different ecological and social evolution from that of chimpanzees, and theirs from us. When the genome of a typical human being has been fully transcribed into our computers, when the same has been done for the average chimpanzee, when the active genes have been extracted from the noise, and when the differences come to be listed, we will have an extraordinary glimpse of the pressures of the Pleistocene era on two different species derived from a common stock. The genes that will be the same will be the genes for basic biochemistry and body planning. Probably the only differences will be in genes for regulating growth and hormonal development. Somehow in their digital language, these genes will tell the foot of a human foetus to grow into a flat object with a heel and a big toe, whereas the same genes in a chimpanzee tell the foot of a chimp foetus to grow into a more curved object with less of a heel and longer, more prehensile toes.

It is mind-boggling even to try to imagine how that can be done – science still has only the vaguest clues about how growth and form are generated by genes – but that it is genes which are responsible there is no doubt. The differences between human beings and chimpanzees are genetic differences and virtually nothing else. Even those who would stress the cultural side of the human condition and deny or doubt the importance of genetic differences between human individuals or races, accept that the differences between us and other species are primarily genetic. Suppose the nucleus of a chimpanzee cell were injected into an enucleated human egg and that egg were implanted into a human womb, and the resulting baby, if it survived to term, were reared in a human family. What would it look like? You do not even need to do the (highly unethical) experiment to know the answer: a chimpanzee. Although it started with human cytoplasm, used a human placenta and had a human upbringing, it would not look even partly human.

Photography provides a helpful analogy. Imagine you take a photograph of a chimpanzee. To develop it you must put it in a bath of developer for the requisite time, but no matter how hard you try, you cannot develop a picture of a human being on the negative by changing the formula of the developer. The genes are the negative; the womb is the developer. Just as a photograph needs to be immersed in a bath of developer before the picture will appear, so the recipe for a chimpanzee, written in digital form in the genes of its egg, needs the correct milieu to become an adult – the nutrients, the fluids, the food and the care – but it already has the information to make a chimpanzee.

The same is not quite true of behaviour. The typical chimpanzee's hardware can be put together in the womb of a foreign species, but its software would be a little awry. A baby chimpanzee would be as socially confused if reared by human beings as Tarzan would be if reared by chimps. Tarzan, for instance, would not learn to speak, and a human-reared chimp would not learn precisely how to appease dominant animals and intimidate subordinates, to make tree nests or to fish for termites. In the case of behaviour, genes are not sufficient, at least in apes.

But they are necessary. If it is mind-boggling to imagine how small differences in linear digital instructions can direct the two per cent difference between a human body and a chimpanzee body, how much more mind-boggling is it to imagine that a few changes in the same instructions can alter the behaviour of a chimpanzee so precisely. I wrote glibly of the mating system of different apes – the promiscuous chimpanzee, the harem-polygamous gorilla and the long-pair-bond human being. In doing so I assumed, even more glibly, that every species behaves in a characteristic way, which, further, assumes that it is somehow at least partly genetically constrained or influenced. How can a bunch of genes, each one a string of quaternary code, make an animal polygamous or monogamous? Answer: I do not have the foggiest idea, but that it can do so I have no doubt. Genes are recipes for both anatomy and behaviour.

History

We've discovered the secret of life.

Francis Crick, 28 February 1953

Though he was only forty-five in 1902, Archibald Garrod was already a pillar of the British medical establishment. He was the son of a knighted professor, the famous Sir Alfred Baring Garrod, whose treatise on that most quintessential of upper-class afflictions, gout, was reckoned a triumph of medical research. His own career was effortlessly distinguished and in due course the inevitable knighthood (for medical work in Malta during the First World War) would be followed by one of the most glittering prizes of all: the Regius professorship of medicine at Oxford in succession to the great Sir William Osler.

You can just picture him, can you not? The sort of crusty and ceremonious Edwardian who stood in the way of scientific progress, stiff in collar, stiff in lip and stiff in mind. You would be wrong. In that year, 1902, Archibald Garrod risked a conjecture that would reveal him to be a man far ahead of his time and somebody who had all but unknowingly put his finger on the answer to the greatest biological mystery of all time: what is a gene? Indeed, so brilliant was his understanding of the gene that he would be long dead before anybody got the point of what he was saying: that a gene was a recipe for a single chemical. What is more, he thought he had found one.

In his work at St Bartholomew's Hospital and Great Ormond Street in London, Garrod had come across a number of patients with a rare and not very serious disease, known as alkaptonuria. Among other more uncomfortable symptoms such as arthritis, their urine and the ear wax turned reddish or inky black on exposure to the air, depending on what they had been eating. In 1901, the parents of one of these patients, a little boy, had a fifth child who also had the affliction. That set Garrod to thinking about whether the problem ran in families. He noticed that the two children's parents were first cousins. So he went back and re-examined the other cases: three of the four families were first-cousin marriages, and of the seventeen alkaptonuria cases he saw, eight were second cousins of each other. But the affliction was not simply passed on from parent to child. Most sufferers had normal children, but the disease could reappear later in their descendants. Luckily, Garrod was abreast of the latest biological thinking. His friend William Bateson was one of those who was excited by the rediscovery just two years before of the experiments of Gregor Mendel, and was writing tomes to popularise and defend the new creed of Mendelism, so Garrod knew he was dealing with a Mendelian recessive – a character that could be carried by one generation but would only be expressed if inherited from both parents. He even used Mendel's botanical terminology, calling such people 'chemical sports'.

This gave Garrod an idea. Perhaps, he thought, the reason that the disease only appeared in those with a double inheritance was because something was missing. Being well versed not only in genetics but also in chemistry, he knew that the black urine and ear wax was caused by a build-up of a substance called homogentisate. Homogentisate might be a normal product of the body's chemistry set, but one that was in most people then broken down and disposed of. The reason for the build-up, Garrod supposed, was because the catalyst that was meant to be breaking down the homogentisate was not working. That catalyst, he thought, must be an enzyme made of protein, and must be the sole product of an inherited factor (or gene, as we would now say). In the afflicted people, the gene produced a defective enzyme; in the carriers this did not matter because the gene inherited from the other parent could compensate.

Thus was born Garrod's bold hypothesis of the 'inborn errors of metabolism', with its far-reaching assumption that genes were there to produce chemical catalysts, one gene to each highly specialised catalyst. Perhaps that was what genes were: devices for making proteins. 'Inborn errors of metabolism', Garrod wrote, 'are due to the failure of a step in the metabolic sequence due to loss or malfunction of an enzyme.' Since enzymes are made of protein, they must be the 'seat of chemical individuality'. Garrod's book, published in 1909, was widely and positively reviewed, but his reviewers comprehensively missed the point. They thought he was talking about rare diseases, not something fundamental to all life. The Garrod theory lay neglected for thirty-five years and had to be rediscovered afresh. By then, genetics was exploding with new ideas and Garrod had been dead for a decade.[1]

We now know that the main purpose of genes is to store the recipe for making proteins. It is proteins that do almost every chemical, structural and regulatory thing that is done in the body: they generate energy, fight infection, digest food, form hair, carry oxygen and so on and on. Every single protein in the body is made from a gene by a translation of the genetic code. The same is not quite true in reverse: there are genes, which are never translated into protein, such as the ribosomal-RNA gene of chromosome 1, but even that is involved in making other proteins. Garrod's conjecture is basically correct: what we inherit from our parents is a gigantic list of recipes for making proteins and for making protein-making machines – and little more.

Garrod's contemporaries may have missed his point, but at least they honoured him. The same could not be said of the man on whose shoulders he stood, Gregor Mendel. You could hardly imagine a more different background from Garrod's than Mendel's. Christened Johann Mendel, he was born in the tiny village of Heinzendorf (now Hynöice) in Northern Moravia in 1822. His father, Anton, was a smallholder who paid his rent in work for his landlord; his health and livelihood were shattered by a falling tree when Johann was sixteen and doing well at the grammar school in Troppau. Anton sold the farm to his son-in-law so he could afford the fees for his son at school and then at university in Olmütz. But it was a struggle and Johann needed a wealthier sponsor, so he became an Augustinian friar, taking the name Brother Gregor. He trundled through theological college in Brünn (now Brno) and emerged a priest. He did a stint as a parish priest, but it was not a success. He tried to become a science teacher after studying at Vienna University, but failed the examination.

Back to Brünn he went, a thirty-one-year-old nonentity, fit only for monastic life. He was good at mathematics and chess playing, had a decent head for figures and possessed a cheerful disposition. He was also a passionate gardener, having learnt from his father how to graft and breed fruit trees. It is here, in the folk knowledge of the peasant culture, that the roots of his insight truly lay. The rudiments of particulate inheritance were dimly understood already by the breeders of cattle and apples, but nobody was being systematic. 'Not one [experiment]', wrote Mendel, 'has been carried out to such an extent and in such a way as to make it possible to determine the number of different forms with certainty according to their separate generations, or definitely to ascertain their statistical relations.' You can hear the audience dozing off already.

So Father Mendel, aged thirty-four, started a series of experiments on peas in the monastery gardens that were to last eight years, involve the planting of over 30,000 different plants – 6,000 in 1860 alone – and eventually change the world forever. Afterwards, he knew what he had done, and published it clearly in the proceedings of the Brünn society for the study of natural science, a journal that found its way to all the best libraries. But recognition never came and Mendel gradually lost interest in the gardens as he rose to become the abbot of Brünn, a kindly, busy and maybe not very pious friar (good food gets more mention in his writing than God). His last years were taken up with an increasingly bitter and lonely campaign against a new tax levied on monasteries by the government, Mendel being the last abbot to pay it. Perhaps his greatest claim to fame, he might have reflected in old age, was that he made Leos Janáček, a talented nineteen-year-old boy in the choir school, the choirmaster of Brünn.

In the garden, Mendel had been hybridising: crossing different varieties of pea plant. But this was no amateur gardener playing at science; this was a massive, systematic and carefully thought-out experiment. Mendel chose seven pairs of varieties of peas to cross. He crossed round-seeded peas with wrinkled ones; yellow cotyledons with green ones; inflated seed pods with wrinkled seed pods; grey seed coats with white seed coats; green unripe pods with yellow unripe pods; axial flowers with terminal flowers; tall stems with dwarf stems. How many more he tried we do not know; all of these not only breed true, but are due to single genes so he must have chosen them knowing already from preliminary work what result to expect. In every case, the resulting hybrids were always like just one parent. The other parent's essence seemed to have vanished. But it had not: Mendel allowed the hybrids to self-fertilise and the essence of the missing grandparent reappeared intact in roughly one-quarter of the cases. He counted and counted – 19,959 plants in the second generation, with the dominant characters outnumbering the recessives by 14,949 to 5,010, or 2.98 to 1. It was, as Sir Ronald Fisher pointed out in the next century, too suspiciously close to a ratio of three. Mendel, remember, was good at mathematics and he knew well before the experiments were over what equation his peas were obeying.[2]

Like a man possessed, Mendel turned from peas to fuschias, maize and other plants. He found the same results. He knew that he had discovered something profound about heredity: characteristics do not mix. There is something hard, indivisible, quantum and particulate at the heart of inheritance. There is no mingling of fluids, no blending of blood; there is instead a temporary joining together of lots of little marbles. In retrospect, this was obvious all along. How else could people account for the fact that a family might contain a child with blue eyes and a child with brown? Darwin, who none the less based his theory on blending inheritance, hinted at the problem several times. 'I have lately been inclined to speculate', he wrote to Huxley in 1857, 'very crudely and indistinctly, that propagation by true fertilisation will turn out to be a sort of mixture, and not true fusion, of two distinct individuals . . . I can understand on no other view the way in which crossed forms go back to so large an extent to ancestral forms.'[3]

Darwin was not a little nervous on the subject. He had recently come under attack from a fierce Scottish professor of engineering, strangely named Fleeming Jenkin, who had pointed out the simple and unassailable fact that natural selection and blending inheritance did not mix. If heredity consisted of blended fluids, then Darwin's theory simply could not work, because each new and advantageous change would be lost in the general dilution of descent. Jenkin illustrated his point with the story of a white man attempting to convert an island of black people to whiteness merely by breeding with them. His white blood would soon be diluted to insignificance. In his heart Darwin knew Jenkin was right, and even the usually ferocious Thomas Henry Huxley was silenced by Jenkin's argument, but Darwin also knew that his own theory was right. He could not square the two. If only he had read Mendel.

Many things are obvious in retrospect, but still take a flash of genius to become plain. Mendel's achievement was to reveal that the only reason most inheritance *seems* to be a blend is because it involves more than one particle. In the early nineteenth century John Dalton had proved that water was actually made up of billions of hard, irreducible little things called atoms and had defeated the rival continuity theorists. So now Mendel had proved the atomic theory of biology. The atoms of biology might have been called all sorts of things: among the names used in the first years of this century were factor, gemmule, plastidule, pangene, biophor, id and idant. But it was 'gene' that stuck.

For four years, starting in 1866, Mendel sent his papers and his ideas to Karl-Wilhelm Nägeli, professor of botany in Munich. With increasing boldness he tried to point out the significance of what he had found. For four years Nägeli missed the point. He wrote back to the persistent monk polite but patronising letters, and told him to try breeding hawkweed. He could not have given more mischievous advice if he tried: hawkweed is apomictic, that is it needs pollen to breed but does not incorporate the genes of the pollinating partner, so cross-breeding experiments give strange results. After struggling with hawkweed Mendel gave up and turned to bees. The results of his extensive experiments on the breeding of bees have never been found. Did he discover their strange 'haplo-diploid' genetics?

Nägeli meanwhile published an immense treatise on heredity that not only failed to mention Mendel's discovery; it also gave a perfect example of it from Nägeli's own work – and still missed the point. Nägeli knew that if you crossed an angora cat with another breed, the angora coat disappeared completely in the next generation, but re-emerged intact in the kittens of the third generation. A clearer example of a Mendelian recessive could hardly be found.

Yet even in his lifetime Mendel came tantalisingly close to full recognition. Charles Darwin, normally so diligent at gleaning ideas from the work of others, even recommended to a friend a book, by W. O. Focke, that contained fourteen different references to Mendel's paper. Yet he seems not to have noticed them himself. Mendel's fate was to be rediscovered, in 1900, long after his own and Darwin's deaths. It happened almost simultaneously in three different places. Each of his rediscoverers – Hugo de Vries, Carl Correns and Erich von Tschermak, all botanists – had laboriously duplicated Mendel's work on different species before he found Mendel's paper.

Mendelism took biology by surprise. Nothing about evolutionary theory demanded that heredity should come in lumps. Indeed, the notion seemed to undermine everything that Darwin had strived to establish. Darwin said that evolution was the accumulation of slight and random changes through selection. If genes were hard things that could emerge intact from a generation in hiding, then how could they change gradually or subtly? In many ways, the early twentieth century saw the triumph of Mendelism over Darwinism. William Bateson expressed the views of many when he hinted that particulate inheritance at least put limits on the power of natural selection. Bateson was a man with a muddled mind and a leaden prose style. He believed that evolution occurred in large leaps from one form to another leaving no intermediates. In pursuit of this eccentric notion, he had published a book in 1894 arguing that inheritance was particulate and had been furiously attacked by 'true' Darwinists ever since. Little wonder he welcomed Mendel with open arms and was the first to translate his papers into English. 'There is nothing in Mendelian discovery which runs counter to the cardinal doctrine that species have arisen [by natural selection]', wrote Bateson, sounding like a theologian claiming to be the true interpreter of St Paul. 'Nevertheless, the result of modern inquiry has unquestionably been to deprive that principle of those supernatural attributes with which it has sometimes been invested ... It cannot in candour be denied that there are passages in the works of Darwin which in some measure give countenance to these abuses of the principle of Natural Selection, but I rest easy in the certainty that had Mendel's paper come into his hands, those passages would have been immediately revised.'[4]

But the very fact that the dreaded Bateson was Mendelism's champion led European evolutionists to be suspicious of it. In Britain, the bitter feud between Mendelians and 'biometricians' persisted for twenty years. As much as anything this passed the torch to the United States where the argument was less polarised. In 1903 an American geneticist called Walter Sutton noticed that chromosomes behave just like Mendelian factors: they come in pairs, one from each parent. Thomas Hunt Morgan, the father of American genetics, promptly became a late convert to Mendelism, so Bateson, who disliked Morgan, gave up being right and fought against the chromosomal theory. By such petty feuds is the history of science often decided. Bateson sank into obscurity while Morgan went on to great things as the founder of a productive school of genetics and the man who lent his name to the unit of genetic distance: the centimorgan. In Britain, it was not until the sharp, mathematical mind of Ronald Fisher was brought to bear upon the matter in 1918 that Darwinism and Mendelism were at last reconciled: far from contradicting Darwin, Mendel had brilliantly vindicated him. 'Mendelism', said Fisher, 'supplied the missing parts of the structure erected by Darwin.'

Yet the problem of mutation remained. Darwinism demanded variety upon which to feed. Mendelism supplied stability instead. If genes were the atoms of biology, then changing them was as heretical as alchemy. The breakthrough came with the first artificial induction of mutation by somebody as different from Garrod and Mendel as could be imagined. Alongside an Edwardian doctor and an Augustinian friar we must place the pugnacious Hermann Joe Muller. Muller was typical of the many brilliant, Jewish scientific refugees crossing the Atlantic in the 1930s in every way except one: he was heading east. A native New Yorker, son of the owner of a small metal-casting business, he had been drawn to genetics at Columbia University, but fell out with his mentor, Morgan, and moved to the University of Texas in 1920. There is a whiff of anti-semitism about Morgan's attitude to the brilliant Muller, but the pattern was all too typical. Muller fought with everybody all his life. In 1932, his marriage on the rocks and his colleagues stealing his ideas (so he said), he attempted suicide, then left Texas for Europe.

Muller's great discovery, for which he was to win the Nobel prize, was that genes are artificially mutable. It was like Ernest Rutherford's discovery a few years before that atomic elements were transmutable and that the word 'atom', meaning in Greek uncuttable, was inappropriate. In 1926, he asked himself, '[Is] mutation unique among biological processes in being itself outside the reach of modification or control, – that it occupies a position similar to that till recently characteristic of atomic transmutation in physical science?'

The following year he answered the question. By bombarding

fruit flies with X-rays, Muller caused their genes to mutate so that their offspring sported new deformities. Mutation, he wrote, 'does not stand as an unreachable god playing its pranks upon us from some impregnable citadel in the germplasm.' Like atoms, Mendel's particles must have some internal structure, too. They could be changed by X-rays. They were still genes after mutation, but not the same genes.

Artificial mutation kick-started modern genetics. Using Muller's X-rays, in 1940 two scientists named George Beadle and Edward Tatum created mutant versions of a bread mould called *Neurospora*. They then worked out that the mutants failed to make a certain chemical because they lacked the working version of a certain enzyme. They proposed a law of biology, which caught on and has proved to be more or less correct: one gene specifies one enzyme. Geneticists began to chant it under their breath: one gene, one enzyme. It was Garrod's old conjecture in modern, biochemical detail. Three years later came Linus Pauling's remarkable deduction that a nasty form of anaemia afflicting mostly black people, in which the red cells turned into sickle shapes, was caused by a fault in the gene for the protein haemoglobin. That fault behaved like a true Mendelian mutation. Things were gradually falling into place: genes were recipes for proteins; mutations were altered proteins made by altered genes.

Muller, meanwhile, was out of the picture. In 1932 his fervent socialism and his equally fervent belief in the selective breeding of human beings, eugenics (he wanted to see children carefully bred with the character of Marx or Lenin, though in later editions of his book he judiciously altered this to Lincoln and Descartes), led him across the Atlantic to Europe. He arrived in Berlin just a few months before Hitler came to power. He watched, horrified, as the Nazis smashed the laboratories of his boss, Oscar Vogt, for not expelling the Jews under his charge.

Muller went east once more, to Leningrad, arriving in the laboratory of Nikolay Vavilov just before the anti-Mendelist Trofim Lysenko caught the ear of Stalin and began his persecution of

Mendelian geneticists in support of his own crackpot theories that wheat plants, like Russian souls, could be trained rather than bred to new regimes; and that those who believed otherwise should not be persuaded, but shot. Vavilov died in prison. Ever hopeful, Muller sent Stalin a copy of his latest eugenic book, but hearing it had not gone down well, found an excuse to get out of the country just in time. He went to the Spanish Civil War, where he worked in the blood bank of the International Brigade, and thence to Edinburgh, arriving with his usual ill luck just in time for the outbreak of the Second World War. He found it hard to do science in a blacked-out Scottish winter wearing gloves in the laboratory and he tried desperately to return to America. But nobody wanted a belligerent, prickly socialist who lectured ineptly and had been living in Soviet Russia. Eventually Indiana University gave him a job. The following year he won the Nobel prize for his discovery of artificial mutation.

But still the gene itself remained an inaccessible and mysterious thing, its ability to specify precise recipes for proteins made all the more baffling by the fact that it must itself be made of protein; nothing else in the cell seemed complicated enough to qualify. True, there was something else in chromosomes: that dull little nucleic acid called DNA. It had first been isolated, from the pus-soaked bandages of wounded soldiers, in the German town of Tübingen in 1869 by a Swiss doctor named Friedrich Miescher. Miescher himself guessed that DNA might be the key to heredity, writing to his uncle in 1892 with amazing prescience that DNA might convey the hereditary message 'just as the words and concepts of all languages can find expression in 24–30 letters of the alphabet'. But DNA had few fans; it was known to be a comparatively monotonous substance: how could it convey a message in just four varieties?[5]

Drawn by the presence of Muller, there arrived in Bloomington, Indiana, a precocious and confident nineteen-year-old, already equipped with a bachelor's degree, named James Watson. He must have seemed an unlikely solution to the gene problem, but the solution he was. Trained at Indiana University by the Italian *émigré* Salvador Luria (predictably, Watson did not hit it off with Muller),

Watson developed an obsessive conviction that genes were made of DNA, not protein. In search of vindication, he went to Denmark, then, dissatisfied with the colleagues he found there, to Cambridge in October 1951. Chance threw him together in the Cavendish laboratory with a mind of equal brilliance captivated by the same conviction about the importance of DNA, Francis Crick.

The rest is history. Crick was the opposite of precocious. Already thirty-five, he still had no PhD (a German bomb had destroyed the apparatus at University College, London, with which he was supposed to have measured the viscosity of hot water under pressure – to his great relief), and his sideways lurch into biology from a stalled career in physics was not, so far, a conspicuous success. He had already fled from the tedium of one Cambridge laboratory where he was employed to measure the viscosity of cells forced to ingest particles, and was busy learning crystallography at the Cavendish. But he did not have the patience to stick to his own problems, or the humility to stick to small questions. His laugh, his confident intelligence and his knack of telling people the answers to their own scientific questions were getting on nerves at the Cavendish. Crick was also vaguely dissatisfied with the prevailing obsession with proteins. The structure of the gene was the big question and DNA, he suspected, was a part of the answer. Lured by Watson, he played truant from his own research to indulge in DNA games. So was born one of the great, amicably competitive and therefore productive collaborations in the history of science: the young, ambitious, supple-minded American who knew some biology and the effortlessly brilliant but unfocused older Briton who knew some physics. It was an exothermic reaction.

Within a few short months, using other people's laboriously gathered but under-analysed facts, they had made possibly the greatest scientific discovery of all time, the structure of DNA. Not even Archimedes leaping from his bath had been granted greater reason to boast, as Francis Crick did in the Eagle pub on 28 February 1953, 'We've discovered the secret of life.' Watson was mortified; he still feared that they might have made a mistake.

But they had not. All was suddenly clear: DNA contained a code written along the length of an elegant, intertwined staircase of a double helix, of potentially infinite length. That code copied itself by means of chemical affinities between its letters and spelt out the recipes for proteins by means of an as yet unknown phrasebook linking DNA to protein. The stunning significance of the structure of DNA was how simple it made everything seem and yet how beautiful. As Richard Dawkins has put it,[6] 'What is truly revolutionary about molecular biology in the post-Watson–Crick era is that it has become digital . . . the machine code of the genes is uncannily computer-like.'

A month after the Watson–Crick structure was published, Britain crowned a new queen and a British expedition conquered Mount Everest on the same day. Apart from a small piece in the *News Chronicle*, the double helix did not make the newspapers. Today most scientists consider it the most momentous discovery of the century, if not the millennium.

Many frustrating years of confusion were to follow the discovery of DNA's structure. The code itself, the language by which the gene expressed itself, stubbornly retained its mystery. Finding the code had been, for Watson and Crick, almost easy – a mixture of guesswork, good physics and inspiration. Cracking the code required true brilliance. It was a four-letter code, obviously: A, C, G and T. And it was translated into the twenty-letter code of amino acids that make up proteins, almost certainly. But how? Where? And by what means?

Most of the best ideas that led to the answer came from Crick, including what he called the adaptor molecule – what we now call transfer RNA. Independently of all evidence, Crick arrived at the conclusion that such a molecule must exist. It duly turned up. But Crick also had an idea that was so good it has been called the greatest wrong theory in history. Crick's 'comma-free' code is more elegant than the one Mother Nature uses. It works like this. Suppose that the code uses three letters in each word (if it uses two, that only gives sixteen combinations, which is too few). Suppose that it has no commas, and nogapsbetweenthewords. Now suppose that it excludes all words that can be misread if you start in the wrong place. So, to take an analogy used by Brian Hayes, imagine all three-letter English words that can be written with the four letters A, S, E and T: ass, ate, eat, sat, sea, see, set, tat, tea and tee. Now eliminate those that can be misread as another word if you start in the wrong place. For example, the phrase ateateat can be misread as 'a tea tea t' or as 'at eat eat' or as 'ate ate at'. Only one of these three words can survive in the code.

Crick did the same with A, C, G and T. He eliminated AAA, CCC, GGG and TTT for a start. He then grouped the remaining sixty words into threes, each group containing the same three letters in the same rotating order. For example, ACT, CTA and TAC are in one group, because C follows A, T follows C, and A follows T in each; while ATC, TCA and CAT are in another group. Only one word in each group survived. Exactly twenty are left – and there are twenty amino acid letters in the protein alphabet! A four-letter code gives a twenty-letter alphabet.

Crick cautioned in vain against taking his idea too seriously. 'The arguments and assumptions which we have had to employ to deduce this code are too precarious for us to feel much confidence in it on purely theoretical grounds. We put it forward because it gives the magic number – twenty – in a neat manner and from reasonable physical postulates.' But the double helix did not have much evidence going for it at first, either. Excitement mounted. For five years everybody assumed it was right.

But the time for theorising was past. In 1961, while everybody else was thinking, Marshall Nirenberg and Johann Matthaei decoded a 'word' of the code by the simple means of making a piece of RNA out of pure U (uracil – the equivalent of DNA's T) and putting it in a solution of amino acids. The ribosomes made a protein by stitching together lots of phenylalanines. The first word of the code had been cracked: UUU means phenylalanine. The comma-free code was wrong, after all. Its great beauty had been that it cannot have what are called reading-shift mutations, in which the loss of one letter makes nonsense of all that follows. Yet the version that Nature has instead chosen, though less elegant, is more tolerant of other kinds of errors. It contains much redundancy with many different three-letter words meaning the same thing.[7]

By 1965 the whole code was known and the age of modern genetics had begun. The pioneering breakthroughs of the 1960s became the routine procedures of the 1990s. And so, in 1995, science could return to Archibald Garrod's long-dead patients with their black urine and say with confidence exactly what spelling mistakes occurred in which gene to cause their alkaptonuria. The story is twentieth-century genetics in miniature. Alkaptonuria, remember, is a very rare and not very dangerous disease, fairly easily treated by dietary advice, so it had lain untouched by science for many years. In 1995, lured by its historical significance, two Spaniards took up the challenge. Using a fungus called *Aspergillus*, they eventually created a mutant that accumulated a purple pigment in the presence of phenylalanine: homogentisate. As Garrod suspected, this mutant had a defective version of the protein called homogentisate dioxygenase. By breaking up the fungal genome with special enzymes, identifying the bits that were different from normal and reading the code therein, they eventually pinned down the gene in question. They then searched through a library of human genes hoping to find one similar enough to stick to the fungal DNA. They found it, on the long arm of chromosome 3, a 'paragraph' of DNA 'letters' that shares fifty-two per cent of its letters with the fungal gene. Fishing out the gene in people with alkaptonuria and comparing it with those who do not have it, reveals that they have just one different letter that counts, either the 690th or the 901st. In each case just a single letter change messes up the protein so it can no longer do its job.[8]

This gene is the epitome of a boring gene, doing a boring chemical job in boring parts of the body, causing a boring disease when broken. Nothing about it is surprising or unique. It cannot be linked with IQ or homosexuality, it tells us nothing about the origin of life, it is not a selfish gene, it does not disobey Mendel's laws, it cannot kill or maim. It is to all intents and purposes exactly the same gene in every creature on the planet – even bread mould has it and uses it for precisely the same job that we do. Yet the gene for homogentisate dioxygenase deserves its little place in history for its story is in microcosm the story of genetics itself. And even this dull little gene now reveals a beauty that would have dazzled Gregor Mendel, because it is a concrete expression of his abstract laws: a story of microscopic, coiled, matching helices that work in pairs, of four-letter codes, and the chemical unity of life.

CHROMOSOME 4

Fate

Open any catalogue of the human genome and you will be con-
fronted not with a list of human potentialities, but a list of diseases,
mostly ones named after pairs of obscure central-European doctors.
This gene causes Niemann–Pick disease; that one causes Wolf–
Hirschhorn syndrome. The impression given is that genes are there
to cause diseases. 'New gene for mental illness', announces a website
on genes that reports the latest news from the front, 'The gene for
early-onset dystonia. Gene for kidney cancer isolated. Autism linked
to serotonin transporter gene. A new Alzheimer's gene. The genetics
of obsessive behaviour.'

   Yet to define genes by the diseases they cause is about as absurd
as defining organs of the body by the diseases they get: livers are
there to cause cirrhosis, hearts to cause heart attacks and brains to
cause strokes. It is a measure, not of our knowledge but of our
ignorance that this is the way the genome catalogues read. It is

literally true that the only thing we know about some genes is that their malfunction causes a particular disease. This is a pitifully small thing to know about a gene, and a terribly misleading one. It leads to the dangerous shorthand that runs as follows: 'X has got the Wolf–Hirschhorn gene.' Wrong. We all have the Wolf–Hirschhorn gene, except, ironically, people who have Wolf–Hirschhorn syndrome. Their sickness is caused by the fact that the gene is missing altogether. In the rest of us, the gene is a positive, not a negative force. The sufferers have the mutation, not the gene.

Wolf–Hirschhorn syndrome is so rare and so serious – its gene is so vital – that its victims die young. Yet the gene, which lies on chromosome 4, is actually the most famous of all the 'disease' genes because of a very different disease associated with it: Huntington's chorea. A mutated version of the gene causes Huntington's chorea; a complete lack of the gene causes Wolf–Hirschhorn syndrome. We know very little about what the gene is there to do in everyday life, but we now know in excruciating detail how and why and where it can go wrong and what the consequence for the body is. The gene contains a single 'word', repeated over and over again: CAG, CAG, CAG, CAG . . . The repetition continues sometimes just six times, sometimes thirty, sometimes more than a hundred times. Your destiny, your sanity and your life hang by the thread of this repetition. If the 'word' is repeated thirty-five times or fewer, you will be fine. Most of us have about ten to fifteen repeats. If the 'word' is repeated thirty-nine times or more, you will in mid-life slowly start to lose your balance, grow steadily more incapable of looking after yourself and die prematurely. The decline begins with a slight deterioration of the intellectual faculties, is followed by jerking limbs and descends into deep depression, occasional hallucination and delusions. There is no appeal: the disease is incurable. But it takes between fifteen and twenty-five horrifying years to run its course. There are few worse fates. Indeed, many of the early psychological symptoms of the disease are just as bad in those who live in an affected family but do not get the disease: the strain and stress of waiting for it to strike are devastating.

The cause is in the genes and nowhere else. Either you have the Huntington's mutation and will get the disease or not. This is determinism, predestination and fate on a scale of which Calvin never dreamed. It seems at first sight to be the ultimate proof that the genes are in charge and that there is nothing we can do about it. It does not matter if you smoke, or take vitamin pills, if you work out or become a couch potato. The age at which the madness will appear depends strictly and implacably on the number of repetitions of the 'word' CAG in one place in one gene. If you have thirty-nine, you have a ninety per cent probability of dementia by the age of seventy-five and will on average get the first symptoms at sixty-six; if forty, on average you will succumb at fifty-nine; if forty-one, at fifty-four; if forty-two, at thirty-seven; and so on until those who have fifty repetitions of the 'word' will lose their minds at roughly twenty-seven years of age. The scale is this: if your chromosomes were long enough to stretch around the equator, the difference between health and insanity would be less than one extra inch.[2]

No horoscope matches this accuracy. No theory of human causality, Freudian, Marxist, Christian or animist, has ever been so precise. No prophet in the Old Testament, no entrail-gazing oracle in ancient Greece, no crystal-ball gipsy clairvoyant on the pier at Bognor Regis ever pretended to tell people exactly when their lives would fall apart, let alone got it right. We are dealing here with a prophecy of terrifying, cruel and inflexible truth. There are a billion three-letter 'words' in your genome. Yet the length of just this one little motif is all that stands between each of us and mental illness.

Huntington's disease, which became notorious when it killed the folk singer Woody Guthrie in 1967, was first diagnosed by a doctor, George Huntington, in 1872 on the eastern tip of Long Island. He noticed that it seemed to run in families. Later work revealed that the Long Island cases were part of a much larger family tree originating in New England. In twelve generations of this pedigree more than a thousand cases of the disease could be found. All were descended from two brothers who emigrated from Suffolk in 1630. Several of their descendants were burnt as witches in Salem in 1693, possibly because of the alarming nature of the disease. But because the mutation only makes itself manifest in middle age, when people have already had children, there is little selective pressure on it to die out naturally. Indeed, in several studies, those with the mutations appear to breed more prolifically than their unaffected siblings.[3]

Huntington's was the first completely dominant human genetic disease to come to light. That means it is not like alkaptonuria in which you must have two copies of the mutant gene, one from each parent, to suffer the symptoms. Just one copy of the mutation will do. The disease seems to be worse if inherited from the father and the mutation tends to grow more severe, by the lengthening of the repeat, in the children of progressively older fathers.

In the late 1970s, a determined woman set out to find the Huntington gene. Following Woody Guthrie's terrible death from the disease, his widow started the Committee to Combat Huntington's Chorea; she was joined by a doctor named Milton Wexler whose wife and three brothers-in-law were suffering from the disease. Wexler's daughter, Nancy, knew she stood a fifty per cent chance of having the mutation herself and she became obsessed with finding the gene. She was told not to bother. The gene would prove impossible to find. It would be like looking for a needle in a haystack the size of America. She should wait a few years until the techniques were better and there was a realistic chance. 'But', she wrote, 'if you have Huntington's disease, you do not have time to wait.' Acting on the report of a Venezuelan doctor, Americo Negrette, in 1979 she flew to Venezuela to visit three rural villages called San Luis, Barranquitas and Laguneta on the shores of Lake Maracaibo. Actually a huge, almost landlocked gulf of the sea, Lake Maracaibo lies in the far west of Venezuela, beyond the Cordillera de Merida.

The area contained a vast, extended family with a high incidence of Huntington's disease. The story they told each other was that the affliction came from an eighteenth-century sailor, and Wexler was able to trace the family tree of the disease back to the early nineteenth century and a woman called, appropriately, Maria Concepcion. She lived in the Pueblos de Agua, villages of houses built on stilts over the water. A fecund ancestor, she had 11,000 descendants in eight generations, 9,000 of whom were still alive in 1981. No less than 371 of them had Huntington's disease when Wexler first visited and 3,600 carried a risk of at least a quarter that they would develop the disease, because at least one grandparent had the symptoms.

Wexler's courage was extraordinary, given that she too might have the mutation. 'It is crushing to look at these exuberant children', she wrote,[4] 'full of hope and expectation, despite poverty, despite illiteracy, despite dangerous and exhausting work for the boys fishing in small boats in the turbulent lake, or for even the tiny girls tending house and caring for ill parents, despite a brutalising disease robbing them of parents, grandparents, aunts, uncles, and cousins – they are joyous and wild with life, until the disease attacks.'

Wexler started searching the haystack. First she collected blood from over 500 people: 'hot, noisy days of drawing blood'. Then she sent it to Jim Gusella's laboratory in Boston. He began to test genetic markers in search of the gene: randomly chosen chunks of DNA, that might or might not turn out to be reliably different in the affected and unaffected people. Fortune smiled on him and by mid-1983 he had not only isolated a marker close to the gene affected, but pinned it down to the tip of the short arm of chromosome 4. He knew which three-millionth of the genome it was in. Home and dry? Not so fast. The gene lay in a region of the text one million 'letters' long. The haystack was smaller, but still vast. Eight years later the gene was still mysterious: 'The task has been arduous in the extreme', wrote Wexler,[4] sounding like a Victorian explorer, 'in this inhospitable terrain at the top of chromosome 4. It has been like crawling up Everest over the past eight years.'

The persistence paid off. In 1993, the gene was found at last, its text was read and the mutation that led to the disease identified. The gene is the recipe for a protein called huntingtin: the protein was discovered after the gene – hence its name. The repetition of the 'word' CAG in the middle of the gene results in a long stretch of glutamines in the middle of the protein (CAG means glutamine in 'genetish'). And, in the case of Huntington's disease, the more gluta-mines there are at this point, the earlier in life the disease begins.[5]

It seems a desperately inadequate explanation of the disease. If the huntingtin gene is damaged, then why does it work all right for the first thirty years of life? Apparently, the mutant form of huntingtin very gradually accumulates in aggregate chunks. Like Alzheimer's disease and BSE, it is this accumulation of a sticky lump of protein within the cell that causes the death of the cell, perhaps because it induces the cell to commit suicide. In Huntington's disease this happens mostly within the brain's dedicated movement-control room, the cerebellum, with the result that movement becomes pro-gressively less easy or controlled.[6]

The most unexpected feature of the stuttering repetition of the word CAG is that it is not confined to Huntington's disease. There are five other neurological diseases caused by so-called 'unstable CAG repeats' in entirely different genes. Cerebellar ataxia is one. There is even a bizarre report that a long CAG repeat deliberately inserted into a random gene in a mouse caused a late-onset, neuro-logical disease rather like Huntington's disease. CAG repeats may therefore cause neurological disease whatever the gene in which they appear. Moreover, there are other diseases of nerve degeneration caused by other stuttering repeats of 'words' and in every case the repeated 'word' begins with C and ends in G. Six different CAG diseases are known. CCG or CGG repeated more than 200 times near the beginning of a gene on the X chromosome causes 'fragile X', a variable but unusually common form of mental retardation (fewer than sixty repeats is normal; up to a thousand is possible). CTG repeated from fifty to one thousand times in a gene on chromosome 19 causes myotonic dystrophy. More than a dozen human diseases are caused by expanded three-letter word repeats – the so-called polyglutamine diseases. In all cases the elongated pro-tein has a tendency to accumulate in indigestible lumps that cause their cells to die. The different symptoms are caused by the fact that different genes are switched on in different parts of the body.[7]

What is so special about the 'word' C*G, apart from the fact that it means glutamine? A clue comes from a phenomenon known as anticipation. It has been known for some time that those with a severe form of Huntington's disease or fragile X are likely to have children in whom the disease is worse or begins earlier than it did in themselves. Anticipation means that the longer the repetition, the longer it is likely to grow when copied for the next generation. We know that these repeats form little loopings of DNA called hairpins. The DNA likes to stick to itself, forming a structure like a hairpin, with the Cs and Gs of the C*G 'words' sticking together across the pin. When the hairpins unfold, the copying mechanism can slip and more copies of the word insert themselves.[8]

A simple analogy might be helpful. If I repeat a word six times in this sentence – cag, cag, cag, cag, cag, cag – you will count it fairly easily. But if I repeat it thirty-six times – cag, cag, cag, cag, cag, cag, cag, cag, cag, cag, cag, cag, cag, cag, cag, cag, cag, cag, cag, cag, cag, cag, cag, cag, cag, cag, cag, cag, cag, cag, cag, cag, cag – I am willing to bet you lose count. So it is with the DNA. The more repeats there are, the more likely the copying mechanism is to insert an extra one. Its finger slips and loses its place in the text. An alternative (or possibly additional) explanation is that the checking system, called mismatch repair, is good at catching small changes, but not big ones in C*G repeats.[9]

This may explain why the disease develops late in life. Laura Mangiarini at Guy's Hospital in London created transgenic mice, equipped with copies of part of the Huntington's gene that contained more than one hundred repeats. As the mice grew older, so the length of the gene increased in all their tissues save one. Up to ten extra CAG 'words' were added to it. The one exception was the cerebellum, the hindbrain responsible for controlling movement. The cells of the cerebellum do not need to change during life once the mice have learnt to walk, so they never divide. It is when cells and genes divide that copying mistakes are made. In human beings, the number of repeats in the cerebellum *falls* during life, though it increases in other tissues. In the cells from which sperm are made, the CAG repeats grow, which explains why there is a relationship between the onset of Huntington's disease and the age of the father: older fathers have sons who get the disease more severely and at a younger age. (Incidentally, it is now known that the mutation rate, throughout the genome, is about five times as high in men as it is in women, because of the repeated replication needed to supply fresh sperm cells throughout life.)[10]

Some families seem to be more prone to the spontaneous appearance of the Huntington's mutation than others. The reason seems to be not only that they have a repeat number just below the threshold (say between twenty-nine and thirty-five), but that it jumps above the threshold about twice as easily as it does in other people with similar repeat numbers. The reason for that is again a simple matter of letters. Compare two people: one has thirty-five CAGs followed by a bunch of CCAs and CCGs. If the reader slips and adds an extra CAG, the repeat number grows by one. The other person has thirty-five CAGs, followed by a CAA then two more CAGs. If the reader slips and misreads the CAA as a CAG, the effect is to add not one but three to the repeat number, because of the two CAGs already waiting.[11]

Though I seem to be getting carried away, and deluging you with details about CAGs in the huntingtin gene, consider: almost none of this was known five years ago. The gene had not been found, the CAG repeat had not been identified, the huntingtin protein was unknown, the link with other neurodegenerative diseases was not even guessed at, the mutation rates and causes were mysterious, the paternal age effect was unexplained. From 1872 to 1993 virtually nothing was known about Huntington's disease except that it was genetic. This mushroom of knowledge has grown up almost overnight since then, a mushroom vast enough to require days in a library merely to catch up. The number of scientists who have published papers on the Huntington's gene since 1993 is close to 100. All about one gene. One of 60,000–80,000 genes in the human genome. If you still need convincing of the immensity of the Pandora's box that James Watson and Francis Crick opened that day in 1953, the Huntington's story will surely persuade you. Compared with the knowledge to be gleaned from the genome, the whole of the rest of biology is but a thimbleful.

And yet not a single case of Huntington's disease has been cured. The knowledge that I celebrate has not even suggested a remedy for the affliction. If anything, in the heartless simplicity of the CAG repeats, it has made the picture look even bleaker for those seeking a cure. There are 100 billion cells in the brain. How can we go in and shorten the CAG repeats in the huntingtin genes of each and every one?

Nancy Wexler relates a story about a woman in the Lake Maracaibo study. She came to Wexler's hut to be tested for neurological signs of the disease. She seemed fine and well but Wexler knew that small hints of Huntington's can be detected by certain tests long before the patient herself sees signs. Sure enough this woman showed such signs. But unlike most people, when the doctors had finished their examination, she asked them what their conclusion was. Did she have the disease? The doctor replied with a question: What do you think? She thought she was all right. The doctors avoided saying what they thought, mentioning the need to get to know people better before they gave diagnoses. As soon as the woman left the room, her friend came rushing in, almost hysterical. What did you tell her? The doctors recounted what they had said. 'Thank God', replied the friend and explained: the woman had said to the friend that she would ask for the diagnosis and if it turned out that she had Huntington's disease, she would immediately go and commit suicide.

There are several things about that story that are disturbing. The first is the falsely happy ending. The woman does have the mutation. She faces a death sentence, whether by her hand or much more slowly. She cannot escape her fate, however nicely she is treated by the experts. And surely the knowledge about her condition is hers to do with as she wishes. If she wishes to act on it and kill herself, who are the doctors to withhold the information? Yet they did the 'right thing', too. Nothing is more sensitive than the results of a test for a fatal disease; telling people the result starkly and coldly may well not be the best thing to do – for them. Testing without counselling is a recipe for misery. But above all the tale drives home the uselessness of diagnosing without curing. The woman thought she was all right. Suppose she had five more years of happy ignorance ahead of her; there is no point in telling her that after that she faces lurching madness.

A person who has watched her mother die from Huntington's disease knows she has a fifty per cent chance of contracting it. But that is not right, is it? No individual can have fifty per cent of this disease. She either has a one hundred per cent chance or zero chance, and the probability of each is equal. So all that a genetic test does is unpackage the risk and tell her whether her ostensible fifty per cent is actually one hundred per cent or is actually zero.

Nancy Wexler fears that science is now in the position of Tiresias, the blind seer of Thebes. By accident Tiresias saw Athena bathing and she struck him blind. Afterwards she repented and, unable to restore his sight, gave him the power of soothsaying. But seeing the future was a terrible fate, since he could see it but not change it. 'It is but sorrow', said Tiresias to Oedipus, 'to be wise when wisdom profits not.' Or as Wexler puts it, 'Do you want to know when you are going to die, especially if you have no power to change the outcome?' Many of those at risk from Huntington's disease, who since 1986 can have themselves tested for the mutation, choose ignorance. Only about twenty per cent of them choose to take the test. Curiously, but perhaps understandably, men are three times as likely to choose ignorance as women. They are more concerned with themselves rather than their progeny.[12]

Even if those at risk choose to know, the ethics are byzantine. If one member of a family takes the test, he or she is in effect testing the whole family. Many parents take the test reluctantly but for the sake of their children. And misconceptions abound, even in textbooks and medical leaflets. Half your children may suffer, says one, addressing parents with the mutation. Not so: each child has a fifty per cent chance, which is very different. How the result of the test is presented is also immensely sensitive. Psychologists have found that people feel better about being told they have a three-quarter chance of an unaffected baby than if they are told they have a one-quarter chance of an affected one. Yet they are the same thing.

Huntington's disease is at the far end of a spectrum of genetics. It is pure fatalism, undiluted by environmental variability. Good living, good medicine, healthy food, loving families or great riches can do nothing about. Your fate is in your genes. Like a pure Augustinian, you go to heaven by God's grace, not by good works. It reminds us that the genome, great book that it is, may give us the bleakest kind of self-knowledge: the knowledge of our destiny, not the kind of knowledge that you can do something about, but the curse of Tiresias.

Yet Nancy Wexler's obsession with finding the gene was driven by her desire to mend it or cure it when she did find it. And she is undoubtedly closer to that goal now than ten years ago. 'I am an optimist', she writes,[4] 'Even though I feel this hiatus in which we will be able only to predict and not to prevent will be exceedingly difficult . . . I believe the knowledge will be worth the risks.'

What of Nancy Wexler herself? Several times in the late 1980s, she and her elder sister Alice sat down with their father Milton to discuss whether either of the women should take the test. The debates were tense, angry and inconclusive. Milton was against taking the test, stressing its uncertainties and the danger of a false diagnosis. Nancy had been determined that she wanted the test, but her determination gradually evaporated in the face of a real possibility. Alice chronicled the discussions in a diary that later became a soul-searching book called *Mapping fate*. The result was that neither woman took the test. Nancy is now the same age as her mother was when she was diagnosed.[13]

CHROMOSOME 5

Environment

Errors, like straws, upon the surface flow;
He who would search for pearls must dive below.
John Dryden, All for Love

It is time for a cold shower. Reader, the author of this book has been misleading you. He has repeatedly used the word 'simple' and burbled on about the surprising simplicity at the heart of genetics. A gene is just a sentence of prose written in a very simple language, he says, preening himself at the metaphor. Such a simple gene on chromosome 3 is the cause, when broken, of alkaptonuria. Another gene on chromosome 4 is the cause, when elongated, of Hunting-ton's chorea. You either have mutations, in which case you get these genetic diseases, or you don't. No need for waffle, statistics or fudge. It is a digital world, this genetics stuff, all particulate inheritance. Your peas are either wrinkled or they are smooth.

You have been misled. The world is not like that. It is a world of greys, of nuances, of qualifiers, of 'it depends'. Mendelian genetics is no more relevant to understanding heredity in the real world than Euclidean geometry is to understanding the shape of an oak tree.

Unless you are unlucky enough to have a rare and serious genetic condition, and most of us do not, the impact of genes upon our lives is a gradual, partial, blended sort of thing. You are not tall or dwarf, like Mendel's pea plants, you are somewhere in between. You are not wrinkled or smooth, but somewhere in between. This comes as no great surprise, because just as we know it is unhelpful to think of water as a lot of little billiard balls called atoms, so it is unhelpful to think of bodies as the products of single, discrete genes. We know in our folk wisdom that genes are messy. There is a hint of your father's looks in your face, but it blends with a hint of your mother's looks, too, and yet is not the same as your sister's – there is something unique about your own looks.

Welcome to pleiotropy and pluralism. Your looks are affected not by a single 'looks' gene, but by lots of them, and by non-genetic factors as well, fashion and free will prominently among them. Chromosome 5 is a good place to start muddying the genetic waters by trying to build a picture that is a little more complicated, a little more subtle and a little more grey than I have painted so far. But I shall not stray too far into this territory yet. I must take things one step at a time, so I will still talk about a disease, though not a very clear-cut one and certainly not a 'genetic' one. Chromosome 5 is the home of several of the leading candidates for the title of the 'asthma gene'. But everything about them screams out pleiotropy – a technical term for multiple effects of multiple genes. Asthma has proved impossible to pin down in the genes. It is maddeningly resistant to being simplified. It remains all things to all people. Almost everybody gets it or some other kind of allergy at some stage in their life. You can support almost any theory about how or why they do so. And there is plenty of room for allowing your political viewpoint to influence your scientific opinion. Those fighting pollution are keen to blame pollution for the increase in asthma. Those who think we have gone soft attribute asthma to central heating and fitted carpets. Those who mistrust compulsory education can lay the blame for asthma at the feet of playground colds. Those who don't like washing their hands can blame excessive hygiene. Asthma, in other words, is much more like real life.

Asthma, moreover, is the tip of an iceberg of 'atopy'. Most asthmatics are also allergic to something. Asthma, eczema, allergy and anaphylaxis are all part of the same syndrome, caused by the same 'mast' cells in the body, alerted and triggered by the same immunoglobulin-E molecules. One person in ten has some form of allergy, the consequences in different people ranging from the mild inconvenience of a bout of hay fever to the sudden and fatal collapse of the whole body caused by a bee sting or a peanut. Whatever factor is invoked to explain the increase in asthma must also be capable of explaining other outbreaks of atopy. In children with a serious allergy to peanuts, if the allergy fades in later life then they are less likely to have asthma.

Yet just about every statement you care to make about asthma can be challenged, including the assertion that it is getting worse. One study asserts that asthma incidence has grown by sixty per cent in the last ten years and that asthma mortality has trebled. Peanut allergy is up by seventy per cent in ten years. Another study, published just a few months later, asserts with equal confidence that the increase is illusory. People are more aware of asthma, more ready to go to the doctor with mild cases, more prepared to define as asthma something that would once have been called a cold. In the 1870s, Armand Trousseau included a chapter on asthma in his *Clinique Médicale*. He described two twin brothers whose asthma was bad in Marseilles and other places but who were cured as soon as they went to Toulon. Trousseau thought this very strange. His emphasis hardly suggests a rare disease. Still, the balance of probability is that asthma and allergy are getting worse and that the cause is, in a word, pollution.

But what kind of pollution? Most of us inhale far less smoke than our ancestors, with their wood fires and poor chimneys, would have done. So it seems unlikely that general smoke can have caused the recent increase. Some modern, synthetic chemicals can cause dramatic and dangerous attacks of asthma. Transported about the countryside in tankers, used in the manufacture of plastics and leaked into the air we breathe, chemicals such as isocyanates, trimetallic anhydride and phthalic anhydride are a new form of pollution and a possible cause of asthma. When one such tanker spilled its load of isocyanate in America it turned the policeman who directed traffic around the wreck into an acute and desperate asthmatic for the remainder of his life. Yet there is a difference between acute, concentrated exposure and the normal levels encountered in everyday life. So far there is no link between low-level exposure to such chemicals and asthma. Indeed, asthma appears in communities that never encounter them. Occupational asthma can be triggered in people who work in much more low-tech, old-fashioned professions, such as grooms, coffee roasters, hairdressers or metal grinders. There are more than 250 defined causes of occupational asthma. By far the commonest asthma trigger – which accounts for about half of all cases – is the droppings of the humble dust mite, a creature that likes our fondness for central-heated indoor winter stuffiness and makes its home inside our carpets and bedding.

The list of asthma triggers given by the American Lung Association covers all walks of life: pollen, feathers, moulds, foods, colds, emotional stress, vigorous exercise, cold air, plastics, metal vapours, wood, car exhaust, cigarette smoke, paint, sprays, aspirin, heart drugs – even, in one kind of asthma, sleep. There is material here for anybody to grind any axe they wish. For instance, asthma is largely an urban problem, as proved by its sudden appearance in places becoming urban for the first time. Jimma, in south-west Ethiopia, is a small city that has sprung up in the last ten years. Its local asthma epidemic is ten years old. Yet the meaning of this fact is uncertain. Urban centres are generally more polluted with car exhaust and ozone, true, but they are also somewhat sanitised.

One theory holds that people who wash themselves as children, or encounter less mud in everyday life, are more likely to become asthmatics: that hygiene, not lack of it, is the problem. Children with elder siblings are less likely to get asthma, perhaps because their siblings bring dirt into the house. In a study of 14,000 children near Bristol, it emerged that those who washed their hands five times a day or more and bathed twice a day, stood a twenty-five per cent chance of having asthma, while those who washed less than three times a day and bathed every other day had slightly over half that risk of asthma. The theory goes that dirt contains bacteria, especially mycobacteria, which stimulate one part of the immune system, whereas routine vaccination stimulates a different part of the immune system. Since these two parts of the immune system (the Th1 cells and the Th2 cells respectively) normally inhibit each other, the modern, sanitised, disinfected and vaccinated child is bequeathed a hyperactive Th2 system, and the Th2 system is specially designed to flush parasites from the wall of the gut with a massive release of histamine. Hence hay fever, asthma and eczema. Our immune systems are set up in such a way that they 'expect' to be educated by soil mycobacteria early in childhood; when they are not, the result is an unbalanced system prone to allergy. In support of this theory, asthmatic attacks can be staved off in mice that have been made allergic to egg-white proteins by the simple remedy of forcing them to inhale mycobacteria. Among Japanese schoolchildren, all of whom receive the BCG inoculation against tuberculosis but only sixty per cent of whom become immune as a result, the immune ones are much less likely to develop allergies and asthma than the non-immune ones. This may imply that giving the Th1 cells some stimulation with a mycobacterial inoculation enables them to suppress the asthmatic effects of their Th2 colleagues. Throw away that bottle steriliser and seek out mycobacteria.[1]

Another, somewhat similar, theory holds that asthma is the unleashed frustration of the worm-fighting element in the immune system. Back in the rural Stone Age (or the Middle Ages, for that matter), the immunoglobulin-E system had its hands full fighting off roundworms, tapeworms, hookworms and flukes. It had no time for being precious about dust mites and cat hairs. Today, it is kept less busy and gets up to mischief instead. This theory rests on a slightly dubious assumption about the ways in which the body's immune system works, but it has quite a lot of support. There is no dose of hay fever that a good tapeworm cannot cure, but then which would you rather have?

Another theory holds that the connection with urbanisation is actually a connection with prosperity. Wealthy people stay indoors, heat their houses and sleep on feather pillows infested with dust mites. Yet another theory is based on the undoubted fact that mild, casual-contact viruses (things like common colds) are increasingly common in societies with rapid transport and compulsory education. Schoolchildren harvest new viruses from the playground at an alarming rate, as every parent knows. When nobody travelled much, the supply of new viruses soon ran out, but today, with parents jetting off to foreign lands or meeting strangers at work all the time, there is an endless supply of new viruses to sample at the saliva-rich, germ-amplifying stations we call primary schools. Over 200 different kinds of virus can cause what is collectively known as the common cold. There is a definite connection between childhood infection with mild viruses, such as respiratory syncitial virus, and asthma susceptibility. The latest vogue theory is that a bacterial infection, which causes non-specific urethritis in women and has been getting commoner at roughly the same rate as asthma, may set up the immune system in such a way that it responds aggressively to allergens in later life. Take your pick. My favourite theory, for what it is worth, is the hygiene hypothesis, though I wouldn't go to the stake for it. The one thing you cannot argue is that asthma is on the increase because 'asthma genes' are on the increase. The genes have not changed that quickly.

So why do so many scientists persist in emphasising that asthma is at least partly a 'genetic disease'? What do they mean? Asthma is a constriction of the airways, which is triggered by histamines, which are in turn released by mast cells, whose transformation is triggered by their immunoglobulin-E proteins, whose activation is caused by the arrival of the very molecule to which they have been sensitised. It is, as biological chains of cause and effect go, a fairly simple concatenation of events. The multiplicity of causes is effected by the design of immunoglobulin E, a protein specially designed to come in many forms, any one of which can fit on to almost any outside molecule or allergen. Although one person's asthma may be triggered by dust mites and another's by coffee beans, the underlying mechanism is still the same: the activation of the immunoglobulin-E system.

Where there are simple chains of biochemical events, there are genes. Every protein in the chain is made by a gene, or, in the case of immunoglobulin E, two genes. Some people are born with, or develop, immunological hair-triggers, presumably because their genes are subtly different from those of other people, thanks to certain mutations.

That much is clear from the fact that asthma tends to run in families (a fact known, incidentally, to the twelfth-century Jewish sage of Cordoba, Maimonides). In some places, by accident of history, asthma mutations are unusually frequent. One such place is the isolated island of Tristan da Cunha, which must have been populated by descendants of an asthma-susceptible person. Despite a fine maritime climate, over twenty per cent of the inhabitants have overt symptoms of asthma. In 1997 a group of geneticists funded by a biotechnology company made the long sea voyage to the island and collected the blood of 270 of the 300 islanders to seek the mutations responsible.

Find those mutant genes and you have found the prime cause of the underlying mechnanism of asthma and with it all sorts of possibilities for a cure. Although hygiene or dust mites can explain why asthma is increasing on average, only differences in genes may explain why one person in a family gets asthma and another does not.

Except, of course, here for the first time we encounter the difficulty with words like 'normal' and 'mutant'. In the case of alkaptonuria it is pretty obvious that one version of the gene is normal and the other one is 'abnormal'. In the case of asthma, it is by no means so obvious. Back in the Stone Age, before feather pillows, an immune system that fired off at dust mites was no handicap, because dust mites were not a pressing problem in a temporary hunting camp on the savannah. And if that same immune system was especially good at killing gut worms, then the theoretical 'asthmatic' was normal and natural; it was the others who were the abnormals and 'mutants' since they had genes that made them more vulnerable to worm infestations. Those with sensitive immunoglobulin-E systems were probably more resistant to worm infestations than those without. One of the dawning realisations of recent decades is just how hard it is to define what is 'normal' and what is mutant.

In the late 1980s, off went various groups of scientists in confident pursuit of the 'asthma gene'. By mid-1998 they had found not one, but fifteen. There were eight candidate genes on chromosome 5 alone, two each on chromosomes 6 and 12, and one on each of chromosomes 11, 13 and 14. This does not even count the fact that two parts of immunoglobulin E, the molecule at the centre of the process, are made by two genes on chromosome 1. The genetics of asthma could be underwritten by all of these genes in varying orders of importance or by any combination of them and others, too.

Each gene has its champion and feelings run high. William Cookson, an Oxford geneticist, has described how his rivals reacted to his discovery of a link between asthma-susceptibility and a marker on chromosome 11. Some were congratulatory. Others rushed into print contradicting him, usually with flawed or small sample sizes. One wrote haughty editorials in medical journals mocking his 'logical disjunctions' and 'Oxfordshire genes'. One or two turned vitriolic in their public criticism and one anonymously accused him of fraud. (To the outside world the sheer nastiness of scientific feuds often comes as something of a surprise; politics, by contrast, is a relatively polite affair.) Things were not improved by a sensational story exaggerating Cookson's discovery in a Sunday newspaper, followed by a television programme attacking the newspaper story and a complaint to the broadcasting regulator by the newspaper. 'After four years of constant scepticism and disbelief', says Cookson mildly,[2] 'we were all feeling very tired.'

This is the reality of gene hunting. There is a tendency among ivory-towered moral philosophers to disparage such scientists as gold-diggers seeking fame and fortune. The whole notion of 'genes for' such things as alcoholism and schizophrenia has been mocked, because such claims have often been later retracted. The retraction is taken not as evidence against that genetic link but as a condemnation of the whole practice of seeking genetic links. And the critics have a point. The simplistic headlines of the press can be very misleading. Yet anybody who gets evidence of a link between a disease and a gene has a duty to publish it. If it proves an illusion, little harm is done. Arguably, more damage has been done by false negatives (true genes that have been prematurely ruled out on inadequate data) than by false positives (suspicions of a link that later prove unfounded).

Cookson and his colleagues eventually got their gene and pinned down a mutation within it that the asthmatics in their sample had more often than others did. It was an asthma gene of sorts. But it only accounted for fifteen per cent of the explanation of asthma and it has proved remarkably hard to replicate the finding in other subjects, a maddening feature of asthma-gene hunting that has recurred with distressing frequency. By 1994 one of Cookson's rivals, David Marsh, was suggesting a strong link between asthma and the gene for interleukin 4, on chromosome 5, based on a study of eleven Amish families. That, too, proved hard to replicate. By 1997 a group of Finns was comprehensively ruling out a connection between asthma and the same gene. That same year a study of a mixed-race population in America concluded that eleven chromosomal regions could be linked to susceptibility to asthma, of which ten were unique to only one racial or ethnic group. In other words, the gene that most defined susceptiblity to asthma in blacks was not the same gene that most defined susceptibility to asthma in whites, which was different again from the gene that most defined susceptibility to asthma in Hispanics.[3]

Gender differences are just as pronounced as racial ones. According to research by the American Lung Association, whereas ozone from petrol-burning cars triggers asthma in men, particulates from diesel engines are more likely to trigger asthma in women. As a rule, males seem to have an early bout of allergy and to outgrow it, while females develop allergies in their mid or late twenties and do not outgrow them (though rules have exceptions, of course, including the rule that rules have exceptions). This could explain something peculiar about asthma inheritance: people often appear to inherit it from allergic mothers, but rarely from their fathers. This could just mean that the father's asthma was long ago in his youth and has been largely forgotten.

The trouble seems to be that there are so many ways of altering the sensitivity of the body to asthma triggers, all along the chain of reactions that leads to the symptoms, that all sorts of genes can be 'asthma genes', yet no single one can explain more than a handful of cases. *ADRB2*, for example, lies on the long arm of chromosome 5. It is the recipe for a protein called the beta-2-adrenergic receptor, which controls bronchodilation and bronchoconstriction – the actual, direct symptom of asthma in the tightening of the windpipe. The commonest anti-asthma drugs work by attacking this receptor. So surely a mutation in *ADRB2* would be a prime 'asthma gene'? The gene was pinned down first in cells derived from the Chinese hamster: a fairly routine 1,239-letter long recipe of DNA. Sure enough a promising spelling difference between some severe nocturnal asthmatics and some non-nocturnal asthmatics soon emerged: letter number 46 was G instead of A. But the result was far from conclusive. Approximately eighty per cent of the nocturnal asthmatics had a G, while fifty-two per cent of the non-nocturnal asthmatics had G. The scientists suggested that this difference was sufficient to prevent the damping down of the allergic system that usually occurs at night.[4]

But nocturnal asthmatics are a small minority. To muddy the waters still further, the very same spelling difference has since been linked to a different asthmatic problem: resistance to asthma drugs. Those with the letter G at the same forty-sixth position in the same gene on both copies of chromosome 5 are more likely to find that their asthma drugs, such as formoterol, gradually become ineffective

over a period of weeks or months than those with a letter A on
both copies.

'More likely' ... 'probably' ... 'in some of': this is hardly the
language of determinism I used for Huntington's disease on chromosome 4. The A to G change at position 46 on the *ADRB2* gene
plainly has something to do with asthma susceptibility, but it cannot
be called the 'asthma gene', nor used to explain why asthma strikes
some people and not others. It is at best a tiny part of the tale,
applicable in a small minority or having a small influence easily
overridden by other factors. You had better get used to such indeterminacy. The more we delve into the genome the less fatalistic it
will seem. Grey indeterminacy, variable causality and vague predisposition are the hallmarks of the system. This is not because what
I said in previous chapters about simple, particulate inheritance is
wrong, but because simplicity piled upon simplicity creates complexity. The genome is as complicated and indeterminate as ordinary
life, because it is ordinary life. This should come as a relief. Simple
determinism, whether of the genetic or environmental kind, is a
depressing prospect for those with a fondness for free will.

CHROMOSOME 6

Intelligence

The hereditarian fallacy is not the simple claim that
IQ is to some degree 'heritable' [but] the equation of
'heritable' with 'inevitable'.   *Stephen Jay Gould*

I have been misleading you, and breaking my own rule into the
bargain. I ought to write it out a hundred times as punishment:
GENES ARE NOT THERE TO CAUSE DISEASES.
Even if a gene causes a disease by being 'broken', most genes are
not 'broken' in any of us, they just come in different flavours. The
blue-eyed gene is not a broken version of the brown-eyed gene, or
the red-haired gene a broken version of the brown-haired gene.
They are, in the jargon, different alleles – alternative versions of the
same genetic 'paragraph', all equally fit, valid and legitimate. They
are all normal; there is no single definition of normality.

Time to stop beating about the bush. Time to plunge headlong
into the most tangled briar of the lot, the roughest, scratchiest, most
impenetrable and least easy of all the brambles in the genetic forest:
the inheritance of intelligence.

Chromosome 6 is the best place to find such a thicket. It was on chromosome 6, towards the end of 1997, that a brave or perhaps foolhardy scientist first announced to the world that he had found a gene 'for intelligence'. Brave, indeed, for however good his evidence, there are plenty of people out there who refuse to admit that such things could exist, let alone do. Their grounds for scepticism are not only a weary suspicion, bred by politically tainted research over many decades, of anybody who even touches the subject of hereditary intelligence, but also a hefty dose of common sense. Mother Nature has plainly not entrusted the determination of our intellectual capacities to the blind fate of a gene or genes; she gave us parents, learning, language, culture and education to program ourselves with.

Yet this is what Robert Plomin announced that he and his colleagues had discovered. A group of especially gifted teenage children, chosen from all over America because they are close to genius in their capacity for schoolwork, are brought together every summer in Iowa. They are twelve- to fourteen-year-olds who have taken exams five years early and come in the top one per cent. They have an IQ of about 160. Plomin's team, reasoning that such children must have the best versions of just about every gene that might influence intelligence, took a blood sample from each of them and went fishing in their blood with little bits of DNA from human chromosome 6. (He chose chromosome 6 because he had a hunch based on some earlier work.) By and by, he found a bit on the long arm of chromosome 6 of the brainboxes which was frequently different from the sequence in other people. Other people had a certain sequence just there, but the clever kids had a slightly different one: not always, but often enough to catch the eye. The sequence lies in the middle of the gene called *IGF2R*.[1]

The history of IQ is not uplifting. Few debates in the history of science have been conducted with such stupidity as the one about intelligence. Many of us, myself included, come to the subject with a mistrustful bias. I do not know what my IQ is. I took a test at school, but was never told the result. Because I did not realise the test was against the clock, I finished little of it and presumably scored low. But then not realising that the test is against the clock does not especially suggest brilliance in itself. The experience left me with little respect for the crudity of measuring people's intelligence with a single number. To be able to measure such a slippery thing in half an hour seems absurd.

Indeed, the early measurement of intelligence was crudely prejudiced in motivation. Francis Galton, who pioneered the study of twins to tease apart innate and acquired talents, made no bones about why he did so:[2]

My general object has been to take note of the varied hereditary faculties of different men, and of the great differences in different families and races, to learn how far history may have shown the practicability of supplanting inefficient human stock by better strains, and to consider whether it might not be our duty to do so by such efforts as may be reasonable, thus exerting ourselves to further the ends of evolution more rapidly and with less distress than if events were left to their own course.

In other words he wanted to selectively cull and breed people as if they were cattle.

But it was in America that intelligence testing turned really nasty. H. H. Goddard took an intelligence test invented by the Frenchman Alfred Binet and applied it to Americans and would-be Americans, concluding with absurd ease that not only were many immigrants to America 'morons', but that they could be identified as such at a glance by trained observers. His IQ tests were ridiculously subjective and biased towards middle-class or western cultural values. How many Polish Jews knew that tennis courts had nets in the middle? He was in no doubt that intelligence was innate:[3] 'the consequent grade of intellectual or mental level for each individual is determined by the kind of chromosomes that come together with the union of the germ cells: that it is but little affected by any later influences except such serious accidents as may destroy part of the mechanism.'

With views like these, Goddard was plainly a crank. Yet he prevailed upon national policy sufficiently to be allowed to test immigrants as they arrived at Ellis Island and was followed by others with even more extreme views. Robert Yerkes persuaded the United States army to let him administer intelligence tests to millions of recruits in the First World War, and although the army largely ignored the results, the experience provided Yerkes and others with the platform and the data to support their claim that intelligence testing could be of commercial and national use in sorting people quickly and easily into different streams. The army tests had great influence in the debate leading to the passage in 1924 by Congress of an Immigration Restriction Act setting strict quotas for southern and eastern Europeans on the grounds that they were stupider than the 'Nordic' types that had dominated the American population prior to 1890. The Act's aims had little to do with science. It was more an expression of racial prejudice and union protectionism. But it found its excuses in the pseudoscience of intelligence testing.

The story of eugenics will be left for a later chapter, but it is little wonder that this history of intelligence testing has left most academics, especially those in the social sciences, with a profound distrust of anything to do with IQ tests. When the pendulum swung away from racism and eugenics just before the Second World War, the very notion of hereditarian intelligence became almost a taboo. People like Yerkes and Goddard had ignored environmental influences on ability so completely that they had tested non-English speakers with English tests and illiterate people with tests requiring them to wield a pencil for the first time. Their belief in heredity was so wishful that later critics generally assumed they had no case at all. Human beings are capable of learning, after all. Their IQ can be influenced by their education so perhaps psychology should start from the assumption that there was no hereditary element at all in intelligence: it is all a matter of training.

Science is supposed to advance by erecting hypotheses and testing them by seeking to falsify them. But it does not. Just as the genetic determinists of the 1920s looked always for confirmation of their ideas and never for falsification, so the environmental determinists of the 1960s looked always for supporting evidence and averted their eyes from contrary evidence, when they should have been actively seeking it. Paradoxically, this is a corner of science where the 'expert' has usually been more wrong than the layman. Ordinary people have always known that education matters, but equally they have always believed in some innate ability. It is the experts who have taken extreme and absurd positions at either end of the spectrum.

There is no accepted definition of intelligence. Is it thinking speed, reasoning ability, memory, vocabulary, mental arithmetic, mental energy or simply the appetite of somebody for intellectual pursuits that marks them out as intelligent? Clever people can be amazingly dense about some things – general knowledge, cunning, avoiding lamp-posts or whatever. A footballer with a poor school record may be able to size up in a split second the opportunity and way to make a telling pass. Music, fluency with language and even the ability to understand other people's minds are capacities and talents that frequently do not seem necessarily to go together. Howard Gardner has argued forcefully for a theory of multiple intelligence that recognises each talent as a separate ability. Robert Sternberg has suggested instead that there are essentially three separate kinds of intelligence – analytic, creative and practical. Analytic problems are ones formulated by other people, clearly defined, that come accompanied by all the information required to solve them, have only one right answer, are disembedded from ordinary experience and have no intrinsic interest: a school exam, in short. Practical problems require you to recognise and formulate the problem itself, are poorly defined, lacking in some relevant information, may or may not have a single answer but spring directly out of everyday life. Brazilian street children who have failed badly at mathematics in school are none the less sophisticated at the kind of mathematics they need in their ordinary lives. IQ is a singularly poor predictor of the ability of professional horse-race handicappers. And some Zambian children are as good at IQ tests that use wire models as they are bad at ones requiring pencil and paper – English children the reverse.

Almost by definition, school concentrates on analytic problems and so do IQ tests. However varied they may be in form and content, IQ tests are inherently biased towards certain kinds of minds. And yet they plainly measure something. If you compare people's performance on different kinds of IQ tests, there is a tendency for them to co-vary. The statistician Charles Spearman first noticed this in 1904 – that a child who does well in one subject tends to do well in others and that, far from being independent, different intelligences do seem well correlated. Spearman called this general intelligence, or, with admirable brevity, 'g'. Some statisticians argue that 'g' is just a statistical quirk – one possible solution among many to the problem of measuring different performances. Others think it is a direct measurement of a piece of folklore: the fact that most people can agree on who is 'clever' and who is not. Yet there is no doubt that 'g' works. It is a better predictor of a child's later performance in school than almost any other measure. There is also some genuinely objective evidence for 'g': the speed with which people perform tasks involving the scanning and retrieval of information correlates with their IQ. And general IQ remains surprisingly constant at different ages: between six and eighteen, your intelligence increases rapidly, of course, but your IQ relative to your peers changes very little. Indeed, the speed with which an infant habituates to a new stimulus correlates quite strongly with later IQ, as if it were almost possible to predict the adult IQ of a baby when only a few months old, assuming certain things about its education. IQ scores correlate strongly with school test results. High-IQ children seem to absorb more of the kind of things that are taught in school.[4]

Not that this justifies fatalism about education: the enormous inter-school and international differences in average achievement at mathematics or other subjects shows how much can still be achieved by teaching. 'Intelligence genes' cannot work in a vacuum; they need environmental stimulation to develop.

So let us accept the plainly foolish definition of intelligence as the thing that is measured by the average of several intelligence tests – 'g' – and see where it gets us. The fact that IQ tests were so crude and bad in the past and are still far from perfect at pinning down something truly objective makes it more remarkable, not less, that they are so consistent. If a correlation between IQ and certain genes shows through what Mark Philpott has called 'the fog of imperfect tests',[5] that makes it all the more likely that there is a strongly heritable element to intelligence. Besides, modern tests have been vastly improved in their objectivity and their insensitivity to cultural background or specific knowledge.

In the heyday of eugenic IQ testing in the 1920s, there was no evidence for heritability of IQ. It was just an assumption of the practitioners. Today, that is no longer the case. The heritability of IQ (whatever IQ is) is a hypothesis that has been tested on two sets of people: twins and adoptees. The results, however you look at them, are startling. No study of the causes of intelligence has failed to find a substantial heritability.

There was a fashion in the 1960s for separating twins at birth, especially when putting them up for adoption. In many cases this was done with no particular thought, but in others it was deliberately done with concealed scientific motives: to test and (it was hoped) demonstrate the prevailing orthodoxy – that upbringing and environment shaped personality and genes did not. The most famous case was that of two New York girls named Beth and Amy, separated at birth by an inquisitive Freudian psychologist. Amy was placed in the family of a poor, overweight, insecure and unloving mother; sure enough, Amy grew up neurotic and introverted, just as Freudian theory would predict. But so – down to the last details – did Beth, whose adoptive mother was rich, relaxed, loving and cheerful. The differences between Amy's and Beth's personalities were almost undetectable when they rediscovered each other twenty years later. Far from demonstrating the power of upbringing to shape our minds, the study proved the very opposite: the power of instinct.[6]

Started by environmental determinists, the study of twins reared apart was later taken up by those on the other side of the argument, in particular Thomas Bouchard of the University of Minnesota. Beginning in 1979, he collected pairs of separated twins from all over the world and reunited them while testing their personalities and IQs. Other studies, meanwhile, concentrated on comparing the IQs of adopted people with those of their adoptive parents and their biological parents or their siblings. Put all such studies together, totting up the IQ tests of tens of thousands of individuals, and the table looks like this. In each case the number is a percentage correlation, one hundred per cent correlation being perfect identity and zero per cent being random difference.

The same person tested twice	87
Identical twins reared together	86
Identical twins reared apart	76
Fraternal twins reared together	55
Biological siblings	47
Parents and children living together	40
Parents and children living apart	31
Adopted children living together	0
Unrelated people living apart	0

Not surprisingly, the highest correlation is between identical twins reared together. Sharing the same genes, the same womb and the same family, they are indistinguishable from the same person taking the test twice. Fraternal twins, who share a womb but are genetically no more similar than two siblings, are much less similar, but they are more similar than ordinary brothers, implying that things experienced in the womb or early family life can matter a little. But the astonishing result is the correlation between the scores of adopted children reared together: zero. Being in the same family has no discernible effect on IQ at all.[7]

The importance of the womb has only recently been appreciated. According to one study, twenty per cent of the similarity in intelligence of a pair of twins can be accounted for by events in the womb, while only five per cent of the intelligence of a pair of siblings can be accounted for by events in the womb. The difference is that twins share the same womb at the same time, whereas siblings do not. The influence upon our intelligence of events that happened in the womb is three times as great as anything our parents did to us after our birth. Thus even that proportion of our intelligence that can be attributed to 'nurture' rather than nature is actually determined by a form of nurture that is immutable and firmly in the past. Nature, on the other hand, continues to express genes throughout youth. It is nature, not nurture, that demands we do not make fatalistic decisions about children's intelligence too young.[8]

This is positively bizarre. It flies in the face of common sense: surely our intelligence is influenced by the books and conversations found in our childhood homes? Yes, but that is not the question. After all, heredity could conceivably account for the fact that both parents and children from the same home like intellectual pursuits. No studies have been done – except for twin and adoption studies – that discriminate between the hereditary and parental-home explanation. The twin and adoption studies are unambiguous at present in favouring the hereditary explanation for the coincidence of parents' and children's IQs. It remains possible that the twin and adoption studies are misleading because they come from too narrow a range of families. These are mostly white, middle-class families, and very few poor or black families are included in the samples. Perhaps it is no surprise that the range of books and conversations found in all middle-class, American, white families is roughly the same. When a study of trans-racial adoptees was done, a small correlation was found between the children's IQ and that of their adoptive parents (nineteen per cent).

But it is still a small effect. The conclusion that all these studies converge upon is that about half of your IQ was inherited, and less than a fifth was due to the environment you shared with your siblings – the family. The rest came from the womb, the school and outside influences such as peer groups. But even this is misleading. Not only does your IQ change with age, but so does its heritability. As you grow up and accumulate experiences, the influence of your genes *increases*. What? Surely, it falls off? No: the heritability of childhood IQ is about forty-five per cent, whereas in late adolescence it rises to seventy-five per cent. As you grow up, you gradually express your own innate intelligence and leave behind the influences stamped on you by others. You select the environments that suit your innate tendencies, rather than adjusting your innate tendencies to the environments you find yourself in. This proves two vital things: that genetic influences are not frozen at conception and that environmental influences are not inexorably cumulative. Heritability does not mean immutability.

Francis Galton, right at the start of this long debate, used an analogy that may be fairly apt. 'Many a person has amused himself', he wrote, 'with throwing bits of stick into a tiny brook and watching their progress; how they are arrested, first by one chance obstacle, then by another; and again, how their onward course is facilitated by a combination of circumstances. He might ascribe much importance to each of these events, and think how largely the destiny of the stick had been governed by a series of trifling accidents. Nevertheless, all the sticks succeed in passing down the current, and in the long run, they travel at nearly the same rate.' So the evidence suggests that intensively exposing children to better tuition has a dramatic effect on their IQ scores, but only temporarily. By the end of elementary school, children who have been in Head Start programmes are no further ahead than children who have not.

If you accept the criticism that these studies mildly exaggerate heritability because they are of families from a single social class, then it follows that heritability will be greater in an egalitarian society than an unequal one. Indeed, the definition of the perfect meritocracy, ironically, is a society in which people's achievements depend on their genes because their environments are equal. We are fast approaching such a state with respect to height: in the past, poor nutrition resulted in many children not reaching their 'genetic' height as adults. Today, with generally better childhood nutrition, more of the differences in height between individuals are due to genes: the heritability of height is, therefore, I suspect, rising. The same cannot yet be said of intelligence with certainty, because environmental variables – such as school quality, family habits, or wealth – may be growing more unequal in some societies, rather than more equal. But it is none the less a paradox: in egalitarian societies, genes matter more.

These heritability estimates apply to the differences between individuals, not those between groups. IQ heritability does seem to be about the same in different populations or races, which might not have been the case. But it is logically false to conclude that because the difference between the IQ of one person and another is approximately fifty per cent heritable, that the difference between the average IQs of blacks and whites or between whites and Asians is due to genes. Indeed, the implication is not only logically false, it so far looks empirically wrong, too. Thus does a large pillar of support for part of the thesis of the recent book *The bell curve*[9] crumble. There are differences between the average IQ scores of blacks and whites, but there is no evidence that these differences are themselves heritable. Indeed, the evidence from cases of cross-racial adoption suggests that the average IQs of blacks reared by and among whites is no different from that of whites.

If IQ is fifty per cent heritable individually, then some genes must influence it. But it is impossible to tell how many. The only thing one can say with certainty is that some of the genes that influence it are variable, that is to say they exist in different versions in different people. Heritability and determinism are very different things. It is entirely possible that the most important genes affecting intelligence are actually non-varying, in which case there would be no heritability for differences caused by those genes, because there would be no such differences For instance, I have five fingers on each hand and so do most people. The reason is that I inherited a genetic recipe that specified five fingers. Yet if I went around the world looking for people with four fingers, about ninety-five per cent of the people I found, possibly more, would be people who had lost fingers in accidents. I would find that having four fingers is something with very low heritability: it is nearly always caused by the environment. But that does not imply that genes had nothing to do with determining finger number. A gene can determine a feature of our bodies that is the same in different people just as surely as it can determine features that are different in different people. Robert Plomin's gene-fishing expeditions for IQ genes will only find genes that come in different varieties, not genes that show no variation. They might therefore miss some important genes.

Plomin's first gene, the *IGF2R* gene on the long arm of chromosome 6, is at first sight an unlikely candidate for an 'intelligence gene'. Its main claim to fame before Plomin linked it with intelligence was that it was associated with liver cancer. It might have been called a 'liver-cancer gene', thus neatly demonstrating the foolishness of identifying genes by the diseases they cause. At some point we may have to decide if its cancer-suppressing function is its main task and its ability to influence intelligence a side-effect, or vice versa. In fact, they could both be side-effects. The function of the protein it encodes is mystifyingly dull: 'the intracellular trafficking of phosphorylated lysosomal enzymes from the Golgi complex and the cell surface to the lysosomes'. It is a molecular delivery van. Not a word about speeding up brain waves.

IGF2R is an enormous gene, with 7,473 letters in all, but the sense-containing message is spread out over a 98,000-letter stretch of the genome, interrupted forty-eight times by nonsense sequences called introns (rather like one of those irritating magazine articles interrupted by forty-eight advertisements). There are repetitive stretches in the middle of the gene that are inclined to vary in length, perhaps affecting the difference between one person's intelligence and another. Since it seems to be a gene vaguely connected with insulin-like proteins and the burning of sugar, it is perhaps relevant that another study has found that people with high IQs are more 'efficient' at using glucose in their brains. While learning to play the computer game called Tetris, high-IQ people show a greater fall in their glucose consumption as they get more practised than do low-IQ people. But this is to clutch at straws. Plomin's gene, if it proves real at all, will be one of many that can influence intelligence in many different ways.[10]

The chief value of Plomin's discovery lies in the fact that, while people may still dismiss the studies of twins and adoptees as too indirect to prove the existence of genetic influences on intelligence, they cannot argue with a direct study of a gene that co-varies with intelligence. One form of the gene is about twice as common in the superintelligent Iowan children as in the rest of the population, a result extremely unlikely to be accidental. But its effect must be small: this version of the gene can only add four points to your IQ, on average. It is emphatically not a 'genius gene'. Plomin hints at up to ten more 'intelligence genes' to come from his Iowa brainboxes. Yet the return of heritable IQ to scientific respectability is greeted with dismay in many quarters. It raises the spectre of eugenic abuse that so disfigured science in the 1920s and 1930s. As Stephen Jay Gould, a severe critic of excessive hereditarianism, has put it: 'A partially inherited low IQ might be subject to extensive improvement through proper education. And it might not. The mere fact of its heritability permits no conclusion.' Indeed. But that is exactly the trouble. It is by no means inevitable that people will react to genetic evidence with fatalism. The discovery of genetic mutations behind conditions like dyslexia has not led teachers to abandon such conditions as incurable – quite the reverse; it has encouraged them to single out dyslexic children for special teaching.[11]

Indeed, the most famous pioneer of intelligence testing, the Frenchman Alfred Binet, argued fervently that its purpose was not to reward gifted children but to give special attention to less gifted ones. Plomin cites himself as a perfect example of the system at work. As the only one of thirty-two cousins from a large family in Chicago to go to college, he credits his fortune to good results on an intelligence test, which persuaded his parents to send him to a more academic school. America's fondness for such tests is in remarkable contrast to Britain's horror of them. The short-lived and notorious eleven-plus exam, predicated on probably-faked data produced by Cyril Burt, was Britain's only mandatory intelligence test. Whereas in Britain the eleven-plus is remembered as a disastrous device that condemned perfectly intelligent children to second-rate schools, in meritocratic America similar tests are the passports to academic success for the gifted but impoverished.

Perhaps the heritability of IQ implies something entirely different, something that once and for all proves that Galton's attempt to discriminate between nature and nurture is misconceived. Consider this apparently fatuous fact. People with high IQs, on average, have more symmetrical ears than people with low IQs. Their whole bodies seem to be more symmetrical: foot breadth, ankle breadth, finger length, wrist breadth and elbow breadth all correlates with IQ.

In the early 1990s there was revived an old interest in bodily symmetry, because of what it can reveal about the body's development during early life. Some asymmetries in the body are consistent: the heart is on the left side of the chest, for example, in most people. But other, smaller asymmetries can go randomly in either direction. In some people the left ear is larger than the right; in others, vice versa. The magnitude of this so-called fluctuating asymmetry is a sensitive measure of how much stress the body was under when developing, stress from infections, toxins or poor nutrition. The fact that people with high IQs have more symmetrical bodies suggests that they were subject to fewer developmental stresses in the womb or in childhood. Or rather, that they were more resistant to such stresses. And the resistance may well be heritable. So the heritability of IQ might not be caused by direct 'genes for intelligence' at all, but by indirect genes for resistance to toxins or infections – genes in other words that work by interacting with the environment. You inherit not your IQ but your ability to develop a high IQ under certain environmental circumstances. How does one parcel that one into nature and nurture? It is frankly impossible.[12]

Support for this idea comes from the so-called Flynn effect. A New Zealand-based political scientist, James Flynn, noticed in the 1980s that IQ is increasing in all countries all the time, at an average rate of about three IQ points per decade. Quite why is hard to determine. It might be for the same reason that height is increasing: improved childhood nutrition. When two Guatemalan villages were given ad-lib protein supplements for several years, the IQ of children, measured ten years later, had risen markedly: a Flynn effect in miniature. But IQ scores are still rising just as rapidly in well-nourished western countries. Nor can school have much to do with it, because interruptions to schooling have demonstrably temporary effects on IQ and because the tests that show the most rapid rises are the ones that have least to do with what is taught in school. It is the ones that test abstract reasoning ability that show the steepest improvements. One scientist, Ulric Neisser, believes that the cause of the Flynn effect is the intense modern saturation of everyday life with sophisticated visual images – cartoons, advertisements, films, posters, graphics and other optical displays – often at the expense of written messages. Children experience a much richer visual environment than once they did, which helps develop their skills in visual puzzles of the kind that dominate IQ tests.[13]

But this environmental effect is, at first sight, hard to square with the twin studies suggesting such a high heritability for IQ. As Flynn himself notes, an increase of fifteen IQ points in five decades implies either that the world was full of dunces in 1950 or that it is full of geniuses today. Since we are not experiencing a cultural renaissance, he concludes that IQ measures nothing innate. But if Neisser is right, then the modern world is an environment that encourages the development of one form of intelligence – facility with visual symbols. This is a blow to 'g', but it does not negate the idea that these different kinds of intelligence are at least partly heritable. After two million years of culture, in which our ancestors passed on learnt local traditions, human brains may have acquired (through natural selection) the ability to find and specialise in those particular skills that the local culture teaches, and that the individual excels in. The environment that a child experiences is as much a consequence of the child's genes as it is of external factors: the child seeks out and creates his or her own environment. If she is of a mechanical bent, she practises mechanical skills; if a bookworm, she seeks out books. The genes may create an appetite, not an aptitude. After all, the high heritability of short-sightedness is accounted for not just by the heritability of eye shape, but by the heritability of literate habits. The heritability of intelligence may therefore be about the genetics of nurture, just as much as the genetics of nature. What a richly satisfying end to the century of argument inaugurated by Galton.

CHROMOSOME 7

Instinct

The tabula of human nature was never rasa.
W. D. Hamilton

Nobody doubts that genes can shape anatomy. The idea that they also shape behaviour takes a lot more swallowing. Yet I hope to persuade you that on chromosome 7 there lies a gene that plays an important part in equipping human beings with an instinct, and an instinct, moreover, that lies at the heart of all human culture.

Instinct is a word applied to animals: the salmon seeking the stream of its birth; the digger wasp repeating the behaviour of its long-dead parents; the swallow migrating south for the winter – these are instincts. Human beings do not have to rely on instinct; they learn instead; they are creative, cultural, conscious creatures. Everything they do is the product of free will, giant brains and brainwashing parents.

So goes the conventional wisdom that has dominated psychology and all other social sciences in the twentieth century. To think otherwise, to believe in innate human behaviour, is to fall into the trap of determinism, and to condemn individual people to a heartless fate written in their genes before they were born. No matter that the social sciences set about reinventing much more alarming forms of determinism to take the place of the genetic form: the parental determinism of Freud; the socio-economic determinism of Marx; the political determinism of Lenin; the peer-pressure cultural determinism of Franz Boas and Margaret Mead; the stimulus–response determinism of John Watson and B. F. Skinner; the linguistic determinism of Edward Sapir and Benjamin Whorf. In one of the great diversions of all time, for nearly a century social scientists managed to persuade thinkers of many kinds that biological causality was determinism while environmental causality preserved free will; and that animals had instincts, but human beings did not.

Between 1950 and 1990 the edifice of environmental determinism came tumbling down. Freudian theory fell the moment lithium first cured a manic depressive, where twenty years of psychoanalysis had failed. (In 1995 a woman sued her former therapist on the grounds that three weeks on Prozac had achieved more than three years of therapy.) Marxism fell when the Berlin wall was built, though it took until the wall came down before some people realised that subservience to an all-powerful state could not be made enjoyable however much propaganda accompanied it. Cultural determinism fell when Margaret Mead's conclusions (that adolescent behaviour was infinitely malleable by culture) were discovered by Derek Freeman to be based on a combination of wishful prejudice, poor data collection and adolescent prank-playing by her informants. Behaviourism fell with a famous 1950s experiment in Wisconsin in which orphan baby monkeys became emotionally attached to cloth models of their mothers even when fed only from wire models, thus refusing to obey the theory that we mammals can be conditioned to prefer the feel of anything that gives us food – a preference for soft mothers is probably innate.[1]

In linguistics, the first crack in the edifice was a book by Noam Chomsky, *Syntactic structures*, which argued that human language, the most blatantly cultural of all our behaviours, owes as much to instinct as it does to culture. Chomsky resurrected an old view of language, which had been described by Darwin as an 'instinctive tendency to acquire an art'. The early psychologist William James, brother of the novelist Henry, was a fervent protagonist of the view that human behaviour showed evidence of more separate instincts than animals, not fewer. But his ideas had been ignored for most of the twentieth century. Chomsky brought them back to life.

By studying the way human beings speak, Chomsky concluded that there were underlying similarities to all languages that bore witness to a universal human grammar. We all know how to use it, though we are rarely conscious of that ability. This must mean that part of the human brain comes equipped by its genes with a specialised ability to learn language. Plainly, the vocabulary could not be innate, or we would all speak one, unvarying language. But perhaps a child, as it acquired the vocabulary of its native society, slotted those words into a set of innate mental rules. Chomsky's evidence for this notion was linguistic: he found regularities in the way we spoke that were never taught by parents and could not be inferred from the examples of everyday speech without great difficulty. For example, in English, to make a sentence into a question we bring the main verb to the front of the statement. But how do we know which verb to bring? Consider the sentence, 'A unicorn that is eating a flower is in the garden.' You can turn that sentence into a question by moving the second 'is' to the front: 'Is a unicorn that is eating a flower in the garden?' But you make no sense if you move the first 'is': 'Is a unicorn that eating a flower is in the garden?' The difference is that the first 'is' is part of a noun phrase, buried in the mental image conjured by not just any unicorn, but any unicorn that is eating a flower. Yet four-year-olds can comfortably use this rule, never having been taught about noun phrases. They just seem to know the rule. And they know it without ever having used or heard the phrase 'a unicorn that is eating a flower' before. That is the beauty of language – almost every statement we make is a novel combination of words.

Chomsky's conjecture has been brilliantly vindicated in the succeeding decades by lines of evidence from many different disciplines.

All converge upon the conclusion that to learn a human language requires, in the words of the psycho-linguist Steven Pinker, a human language instinct. Pinker (who has been called the first linguist capable of writing readable prose) persuasively gathered the strands of evidence for the innateness of language skills. There is first the universality of language. All human people speak languages of comparable grammatical complexity, even those isolated in the highlands of New Guinea since the Stone Age. All people are as consistent and careful in following implicit grammatical rules, even those without education and who speak what are patronisingly thought to be 'slang' dialects. The rules of inner-city black Ebonics are just as rational as the rules of the Queen's English. To prefer one to another is mere prejudice. For example, to use double negatives ('Don't nobody do this to me . . .') is considered proper in French, but slang in English. The rule is just as consistently followed in each.

Second, if these rules were learnt by imitation like the vocabulary, then why would four-year-olds who have been happily using the word 'went' for a year or so, suddenly start saying 'goed'? The truth is that although we must teach our children to read and write – skills for which there is no specialised instinct – they learn to speak by themselves at a much younger age with the least of help from us. No parent uses the word 'goed', yet most children do at some time. No parent explains that the word 'cup' refers to all cup-like objects, not this one particular cup, nor just its handle, nor the material from which it is made, nor the action of pointing to a cup, nor the abstract concept of cupness, nor the size or temperature of cups. A computer that was required to learn language would have to be laboriously equipped with a program that ignored all these foolish options – with an instinct, in other words. Children come pre-programmed, innately constrained to make only certain kinds of guess.

But the most startling evidence for a language instinct comes from a series of natural experiments in which children imposed grammatical rules upon languages that lacked them. In the most famous case, studied by Derek Bickerton, a group of foreign labourers brought together on Hawaii in the nineteenth century developed a pidgin language – a mixture of words and phrases whereby they could communicate with each other. Like most such pidgins, the language lacked consistent grammatical rules and remained both laboriously complex in the way it had to express things and relatively simple in what it could express. But all that changed when for the first time a generation of children learnt the language in their youth. The pidgin acquired rules of inflection, word order and grammar that made it a far more efficient and effective language – a creole. In short, as Bickerton concluded, pidgins become creoles only after they are learnt by a generation of children, who bring instinct to bear on their transformation.

Bickerton's hypothesis has received remarkable support from the study of sign language. In one case, in Nicaragua, special schools for the deaf, established for the first time in the 1980s, led to the invention, *de novo*, of a whole new language. The schools taught lip-reading with little success, but in the playground the children brought together the various hand signs they used at home and established a crude pidgin language. Within a few years, as younger children learnt this pidgin, it was transformed into a true sign language with all the complexity, economy, efficiency and grammar of a spoken language. Once again, it was children who made the language, a fact that seems to suggest that the language instinct is one that is switched off as the child reaches adulthood. This accounts for our difficulty in learning new languages, or even new accents, as adults. We no longer have the instinct. (It also explains why it is so much harder, even for a child, to learn French in a classroom than on holiday in France: the instinct works on speech that it hears, not rules that it memorises.) A sensitive period during which something can be learnt, and outside which it cannot, is a feature of many animals' instincts. For instance, a chaffinch will only learn the true song of its species if exposed to examples between certain ages. That the same is true of human beings was proved in a brutal way by the true story of Genie, a girl discovered in a Los Angeles apartment aged thirteen. She had been kept in a single sparsely furnished room all her life and deprived of almost all human contact.

She had learnt two words, 'Stopit' and 'Nomore'. After her release from this hell she rapidly acquired a larger vocabulary, but she never learnt to handle grammar – she had passed the sensitive period when the instinct is expressed.

Yet even bad ideas take a lot of killing, and the notion that language is a form of culture that can shape the brain, rather than vice versa, has been an inordinate time a-dying. Even though the canonical case histories, like the lack of a concept of time in the Hopi language and hence in Hopi thought, have been exposed as simple frauds, the notion that language is a cause rather than consequence of the human brain's wiring survives in many social sciences. It would be absurd to argue that only Germans can understand the concept of taking pleasure at another's misfortune; and that the rest of us, not having a word for *Schadenfreude*, find the concept entirely foreign.[2]

Further evidence for the language instinct comes from many sources, not least from detailed studies of the ways in which children develop language in their second year of life. Irrespective of how much they are spoken to directly, or coached in the use of words, children develop language skills in a predictable order and pattern. And the tendency to develop language late has been demonstrated by twin studies to be highly heritable. Yet for many people the most persuasive evidence for the language instinct comes from the hard sciences: neurology and genetics. It is hard to argue with stroke victims and real genes. The same part of the brain is consistently used for language processing (in most people, on the left side of the brain), even the deaf who 'speak' with their hands, though sign language also uses part of the right hemisphere.[3]

If a particular one of these parts of the brain is damaged, the effect is known as Broca's aphasia, an inability to use or understand all but the simplest grammar, even though the ability to understand sense remains unaffected. For instance, a Broca's aphasic can easily answer questions such as 'Do you use a hammer for cutting?' but has great difficulty with: 'The lion was killed by the tiger. Which one is dead?' The second question requires sensitivity to the grammar encoded in word order, which is known by just this one part of the brain. Damage to another area, Wernicke's area, has almost the opposite effect – people with such damage produce a rich but senseless stream of words. It appears as if Broca's area generates speech and Wernicke's area instructs Broca's area what speech to generate. This is not the whole story, for there are other areas active in language processing, notably the insula (which may be the region that malfunctions in dyslexia).[4]

There are two genetic conditions that affect linguistic ability. One is Williams syndrome, caused by a change in a gene on chromosome 11, in which affected children are very low in general intelligence, but have a vivid, rich and loquacious addiction to using language. They chatter on, using long words, long sentences and elaborate syntax. If asked to refer to an animal, they are as likely to choose something bizarre like an aardvark as a cat or a dog. They have a heightened ability to learn language but at the expense of sense: they are severely mentally retarded. Their existence seems to undermine the notion, which most of us have at one time or another considered, that reason is a form of silent language.

The other genetic condition has the opposite effect: it lowers linguistic ability without apparently affecting intelligence, or at least not consistently. Known as specific language impairment (SLI), this condition is at the centre of a fierce scientific fight. It is a battleground between the new science of evolutionary psychology and the old social sciences, between genetic explanations of behaviour and environmental ones. And the gene is here on chromosome 7.

That the gene exists is not at issue. Careful analysis of twin studies unambiguously points to a strong heritability for specific language impairment. The condition is not associated with neurological damage during birth, is not associated with linguistically impoverished upbringings, and is not caused by general mental retardation. According to some tests – and depending on how it is defined – the heritability approaches one hundred per cent. That is, identical twins are twice as likely to share the condition as fraternal twins.[5]

That the gene in question is on chromosome 7 is also not in much doubt. In 1997 a team of Oxford-based scientists pinned down a genetic marker on the long arm of chromosome 7, one form of which co-occurs with the condition of SLI. The evidence, though based only on one large English family, was strong and unambiguous.[6]

So why the battleground? The argument rages about what SLI is. To some it is merely a general problem with the brain that affects many aspects of language-producing ability, including principally the ability to articulate words in the mouth and to hear sounds correctly in the ear. The difficulty the subjects experience with language follow from these sensory problems, according to this theory. To others, this is highly misleading. The sensory and voice problems exist, to be sure, in many victims of the condition, but so does something altogether more intriguing: a genuine problem understanding and using grammar that is quite independent of any sensory deficits. The only thing both sides can agree upon is that it is thoroughly disgraceful, simplistic and sensationalist of the media to portray this gene, as they have done, as a 'grammar gene'.

The story centres on a large English family known as the Ks. There are three generations. A woman with the condition married an unaffected man and had four daughters and one son: all save one daughter were affected and they in turn had between them twenty-four children, ten of whom have the condition. This family has got to know the psychologists well; rival teams besiege them with a battery of tests. It is their blood that led the Oxford team to the gene on chromosome 7. The Oxford team, working with the Institute of Child Health in London, belongs to the 'broad' school of SLI, which argues that the grammar-deficient skills of the K family members stem from their problems with speech and hearing. Their principal opponent and the leading advocate of the 'grammar theory' is a Canadian linguist named Myrna Gopnik.

In 1990 Gopnik first suggested that the K family and others like them have a problem knowing the basic rules of English grammar. It is not that they cannot know the rules, but that they must learn them consciously and by heart, rather than instinctively internalise them. For example, if Gopnik shows somebody a cartoon of an imaginary creature and with it the words 'This is a Wug', then shows them a picture of two such creatures together with the words 'These are . . .', most people reply, quick as a flash, 'Wugs'. Those with SLI rarely do so, and if they do, it is after careful thought. The English plural rule, that you add an 's' to the end of most words, is one they seem not to know. This does not prevent those with SLI knowing the plural of most words, but they are stumped by novel words that they have not seen before, and they make the mistake of adding 's' to fictitious words that the rest of us would not, such as 'saess'. Gopnik hypothesises that they store English plurals in their minds as separate lexical entries, in the same way that we all store singulars. They do not store the grammatical rule.[7]

The problem is not, of course, confined to plurals. The past tense, the passive voice, various word-order rules, suffixes, word-combination rules and all the laws of English we each so unconsciously know, give SLI people difficulty, too. When Gopnik first published these findings, after studying the English family, she was immediately and fiercely attacked. It was far more reasonable, said one critic, to conclude that the source of the variable performance problems lay in the language-processing system, rather than the underlying grammar. Grammatical forms like plural and past tense were particularly vulnerable, in English, in individuals with speech defects. It was misleading of Gopnik, said another pair of critics, to neglect to report that the K family has a severe congenital speech disorder, which impairs their words, phonemes, vocabulary and semantic ability as well as their syntax. They had difficulty understanding many other forms of syntactical structure such as reversible passives, post-modified subjects, relative clauses and embedded forms.[8]

These criticisms had a whiff of territoriality about them. The family was not Gopnik's discovery: how dare she assert novel things about them? Moreover, there was some support for her idea in at least part of the criticism: that the disorder applied to all syntactical forms. And to argue that the grammatical difficulty must be caused by the mis-speaking problem, because mis-speaking goes with the grammatical difficulty, was circular.

Gopnik was not one to give up. She broadened the study to Greek and Japanese people as well, using them for various ingenious experiments designed to show the same phenomena. For example, in Greek, the word 'likos' means wolf. The word 'likanthropos' means wolfman. The word 'lik', the root of wolf, never appears on its own. Yet most Greek speakers automatically know that they must drop the '-os' to find the root if they wish to combine it with another word that begins with a vowel, like '-anthropos', or drop only the 's', to make 'liko-' if they wish to combine it with a word that begins with a consonant. It sounds a complicated rule, but even to English speakers it is immediately familiar: as Gopnik points out, we use it all the time in new English words like 'technophobia'.

Greek people with SLI cannot manage the rule. They can learn a word like 'likophobia' or 'likanthropos', but they are very bad at recognising that such words have complex structures, built up from different roots and suffixes. As a result, to compensate, they effectively need a larger vocabulary than other people. 'You have to think of them', says Gopnik, 'as people without a native language.' They learn their own tongue in the same laborious way that we, as adults, learn a foreign language, consciously imbibing the rules and words.[9]

Gopnik acknowledges that some SLI people have low IQ on non-verbal tests, but on the other hand some have above-average IQ. In one pair of fraternal twins, the SLI one had higher non-verbal IQ than the unaffected twin. Gopnik also acknowledges that most SLI people have problems speaking and hearing as well, but she contends that by no means all do and that the coincidence is irrelevant. For instance, people with SLI have no trouble learning the difference between 'ball' and 'bell', yet they frequently say 'fall' when they mean 'fell' – a grammatical, not a vocabulary difference. Likewise, they have no difficulty discerning the difference between rhyming words, like 'nose' and 'rose'. Gopnik was furious when one of her opponents described the K family members' speech as 'unintelligible' to outsiders. Having spent many hours with them, talking, eating pizza and attending family celebrations, she says they are perfectly comprehensible. To prove the irrelevance of speaking and hearing difficulties, she has devised written tests, too. For example, consider the following pair of sentences: 'He was very happy last week when he was first.' 'He was very happy last week when he is first.' Most people immediately recognise that the first is grammatical and the second is not. SLI people think they are both acceptable statements. It is hard to conceive how this could be due to a hearing or speaking difficulty.[10]

None the less, the speaking-and-hearing theorists have not given up. They have recently shown that SLI people have problems with 'sound masking', whereby they fail to notice a pure tone when it is masked by preceding or following noise, unless the tone is forty-five decibels more intense than is detectable to other people. In other words, SLI people have more trouble picking out the subtler sounds of speech from the stream of louder sounds, so they might, for example, miss the '-ed' on the end of a word.

But instead of supporting the view that this explains the entire range of SLI symptoms, including the difficulty with grammatical rules, this lends credence to a much more interesting, evolutionary explanation: that the speech and hearing parts of the brain are next door to the grammar parts and both are damaged by SLI. SLI results from damage to the brain caused in the third trimester of pregnancy by an unusual version of a gene on chromosome 7. Magnetic-resonance imaging confirms the existence of the brain lesion and the rough location. It occurs, not surprisingly, in one of the two areas devoted to speech and language processing, the areas known as Broca's and Wernicke's areas.

There are two areas in the brains of monkeys that correspond precisely to these areas. The Broca-homologue is used for controlling the muscles of the monkey's face, larynx, tongue and mouth. The Wernicke-homologue is used for recognising sound sequences and the calls of other monkeys. These are exactly the non-linguistic problems that many SLI people have: controlling facial muscles and hearing sounds distinctly. In other words, when ancestral human beings first evolved a language instinct, it grew in the region devoted to sound production and processing. That sound-production and processing module remained, with its connections to facial muscles and ears, but the language instinct module grew on top of it, with its innate capacity for imposing the rules of grammar on the vocabulary of sounds used by members of the species. Thus, although no other primate can learn grammatical language at all – and we are indebted to many diligent, sometimes gullible and certainly wishful trainers of chimpanzees and gorillas for thoroughly exhausting all possibilities to the contrary – language is intimately physically connected with sound production and processing. (Yet not too intimately: deaf people redirect the input and output of the language module to the eyes and hands respectively.) A genetic lesion in that part of the brain therefore affects grammatical ability, speech and hearing – all three modules.[11]

No better proof could be adduced for William James's nineteenth-century conjecture that human beings evolved their complex behaviour by adding instincts to those of their ancestors, not by replacing instincts with learning. James's theory was resurrected in the late 1980s by a group of scientists calling themselves evolutionary psychologists. Prominent among them were the anthropologist John Tooby, the psychologist Leda Cosmides and the psycho-linguist Steven Pinker. Their argument, in a nutshell, is this. The main goal of twentieth-century social science has been to trace the ways in which our behaviour is influenced by the social environment; instead, we could turn the problem on its head and trace the ways in which the social environment is the product of our innate social instincts. Thus the fact that all people smile at happiness and frown when worried, or that men from all cultures find youthful features sexually attractive in women, may be expressions of instinct, not culture. Or the universality of romantic love and religious belief might imply that these are influenced by instinct more than tradition. Culture, Tooby and Cosmides hypothesised, is the product of individual psychology more than vice versa. Moreover, it has been a gigantic mistake to oppose nature to nurture, because all learning depends on innate capacities to learn and innate constraints upon what is learnt. For instance, it is much easier to teach a monkey (and a man) to fear snakes than it is to teach it to fear flowers. But you still have to teach it. Fear of snakes is an instinct that has to be learnt.[12]

The 'evolutionary' in evolutionary psychology refers not so much to an interest in descent with modification, nor to the process of natural selection itself – interesting though these are, they are inaccessible to modern study in the case of the human mind, because they happen too slowly – but to the third feature of the Darwinian paradigm: the concept of adaptation. Complex biological organs can be reverse-engineered to discern what they are 'designed' to do, in just the same way that sophisticated machines can be so studied. Steven Pinker is fond of pulling from his pocket a complicated thing designed for pitting olives to explain the process of reverse engineering. Leda Cosmides prefers a Swiss-army knife to make a similar point. In each case, the machines are meaningless except when described in terms of their particular function: what is this blade for? It would be meaningless to describe the working of a camera without reference to the fact that it is designed for the making of images. In the same way, it is meaningless to describe the human (or animal) eye without mentioning that it is specifically designed for approximately the same purpose.

Pinker and Cosmides both contend that the same applies to the human brain. Its modules, like the different blades of a Swiss-army knife, are most probably designed for particular functions. The alternative, that the brain is equipped with random complexity, from which its different functions fall out as fortunate by-products of the physics of complexity – an idea still favoured by Chomsky – defies all evidence. There is simply nothing to support the conjecture that the more detailed you make a network of microprocessors, the more functions they will acquire. Indeed, the 'connectionist' approach to neural networks, largely misled by the image of the brain as a general-purpose network of neurons and synapses, has tested the idea thoroughly and found it wanting. Pre-programmed design is required for the solving of pre-ordained problems.

There is a particular historical irony here. The concept of design in nature was once one of the strongest arguments advanced against evolution. Indeed, it was the argument from design that kept evolutionary ideas at bay throughout the first half of the nineteenth century. Its most able exponent, William Paley, famously observed that if you found a stone on the ground, you could conclude little of interest about how it got there. But if you found a watch, you would be forced to conclude that somewhere there was a watchmaker. Thus the exquisite, functional design apparent in living creatures was manifest evidence for God. It was Darwin's genius to use the argument from design just as explicitly but in the service of the opposite conclusion: to show that Paley was wrong. A 'blind watchmaker' (in Richard Dawkins's phrase) called natural selection, acting step by step on the natural variation in the creature's body, over many millions of years and many millions of individuals, could just as easily account for complex adaptation. So successfully has Darwin's hypothesis been supported that complex adaptation is now considered the primary evidence that natural selection has been at work.[13]

The language instinct that we all possess is plainly one such complex adaptation, beautifully designed for clear and sophisticated communication between individuals. It is easy to conceive how it was advantageous for our ancestors on the plains of Africa to share detailed and precise information with each other at a level of sophistication unavailable to other species. 'Go a short way up that valley and turn left by the tree in front of the pond and you will find the giraffe carcass we just killed. Avoid the brush on the right of the tree that is in fruit, because we saw a lion go in there.' Two sentences pregnant with survival value to the recipient; two tickets for success in the natural-selection lottery, yet wholly incomprehensible without a capacity for understanding grammar, and lots of it.

The evidence that grammar is innate is overwhelming and diverse. The evidence that a gene somewhere on chromosome 7 usually plays a part in building that instinct in the developing foetus's brain is good, though we have no idea how large a part that gene plays. Yet most social scientists remain fervently resistant to the idea of genes whose primary effect seems to be to achieve the development of grammar directly. As is clear in the case of the gene on chromosome 7, many social scientists prefer to argue, despite much evidence, that the gene's effects on language are mere side-effects of its direct effect on the ability of the brain to understand speech. After a century in which the dominating paradigm has been that instincts are confined to 'animals' and are absent from human beings, this reluctance is not surprising. This whole paradigm collapses once you consider the Jamesian idea that some instincts cannot develop without learnt, outside inputs.

This chapter has followed the arguments of evolutionary psychology, the reverse-engineering of human behaviour to try to understand what particular problems it was selected to solve. Evolutionary psychology is a new and remarkably successful discipline that has brought sweeping new insights to the study of human behaviour in many fields. Behaviour genetics, which was the subject of the chapter on chromosome 6, aims at roughly the same goal. But the approach to the subject is so different that behaviour genetics and evolutionary psychology are embarked on a collision course. The problem is this: behaviour genetics seeks variation between individuals and seeks to link that variation to genes. Evolutionary psychology seeks common human behaviour – human universals, features found in every one of us – and seeks to understand how and why such behaviour must have become partly instinctive. It therefore assumes no individual differences exist, at least for important behaviours. This is because natural selection consumes variation: that is its job. If one version of a gene is much better than another, then the better version will soon be universal to the species and the worse version will soon be extinct. Therefore, evolutionary psychology concludes that if behaviour geneticists find a gene with common variation in it, then it may not be a very important gene, merely an auxiliary. Behaviour geneticists retort that every human gene yet investigated turns out to have variants, so there must be something wrong with the argument from evolutionary psychology.

In practice, it may gradually emerge that the disagreement between these two approaches is exaggerated. One studies the genetics of common, universal, species-specific features. The other studies the genetics of individual differences. Both are a sort of truth. All human beings have a language instinct, whereas all monkeys do not, but that instinct does not develop equally well in all people. Somebody with SLI is still far more capable of learning language than Washoe, Koko, Nim or any of the other trained chimpanzees and gorillas.

The conclusions of both behaviour genetics and evolutionary psychology remain distinctly unpalatable to many non-scientists, whose main objection is a superficially reasonable argument from incredulity. How can a gene, a stretch of DNA 'letters', cause a behaviour? What conceivable mechanism could link a recipe for a protein with an ability to learn the rule for making the past tense in English? I admit that this seems at first sight a mighty leap, requiring more faith than reason. But it need not be, because the genetics of behaviour is, at root, no different from the genetics of embryonic development. Suppose that each module of the brain grows its adult form by reference to a series of chemical gradients laid down in the developing embryo's head – a sort of chemical road map for neurons. Those chemical gradients could themselves be the product of genetic mechanisms. Hard though it is to imagine genes and proteins that can tell exactly where they are in the embryo, there is no doubting they exist. As I shall reveal when discussing chromosome 12, such genes are one of the most exciting products of modern genetic research. The idea of genes for behaviour is no more strange than the idea of genes for development. Both are mind-boggling, but nature has never found human incomprehension a reason for changing her methods.

Conflict

Xq28 – Thanks for the genes mom.
T shirt sold in gay and lesbian
bookstores in the mid-1990s

A detour into linguistics has brought us face to face with the startling implications of evolutionary psychology. If it has left you with an unsettling feeling that something else is in control, that your own abilities, linguistic and psychological, were somewhat more instinctively determined than you proudly imagined, then things are about to get a lot worse. The story of this chapter is perhaps the most unexpected in the whole history of genetics. We have got used to thinking of genes as recipes, passively awaiting transcription at the discretion of the collective needs of the whole organism: genes as servants of the body. Here we encounter a different reality. The body is the victim, plaything, battleground and vehicle for the ambitions of genes.

The next largest chromosome after number seven, is called the X chromosome. X is the odd one out, the misfit. Its pair, the chromosome with which it has some affinity of sequence, is not, as in every other case, an identical chromosome, but is the Y chromosome, a tiny and almost inert stub of a genetic afterthought. At least that is the case in male mammals and flies, and in female butterflies and birds. In female mammals or male birds there are instead two X chromosomes, but they are still somewhat eccentric. In every cell in the body, instead of both expressing their genetic message at equal volume, one of the two at random packs itself up into a tight bundle known as a Barr body and remains inert.

The X and Y chromosomes are known as the sex chromosomes for the obvious reason that they determine, with almost perfect predestination, the sex of the body. Everybody gets an X chromosome from his or her mother. But if you inherited a Y chromosome from your father, you are a man; if you inherited an X chromosome from your father, you are a woman. There are rare exceptions, superficially female people with an X and a Y, but they are exceptions that prove the rule. The key masculinising gene on the Y chromosome is missing or broken in such people.

Most people know this. It does not take much exposure to school biology to come across the X and Y chromosomes. Most people also know that the reason colour-blindness, haemophilia and some other disorders are much more common in men is that these genes are on the X chromosome. Since men have no 'spare' X chromosome, they are much more likely to suffer from these recessive problems than women – as one biologist has put it, the genes on the X chromosome fly without co-pilots in men. But there are things about the X and Y chromosomes most people do not know, disturbing, strange things that have unsettled the very foundations of biology.

It is not often that you find language like this in one of the most sober and serious of all scientific publications, the *Philosophical Transactions of the Royal Society*: 'The mammalian Y chromosome is thus likely to be engaged in a battle in which it is outgunned by its opponent. A logical consequence is that the Y should run away and hide, shedding any transcribed sequences that are not essential to its function.'[1] 'A battle', 'outgunned', 'opponent', 'run away'? These are hardly the sort of things we can expect molecules of DNA to do. Yet the same language, a little more technically phrased, appears in another scientific paper about the Y chromosome, entitled 'The enemies within: intergenomic conflict, interlocus contest evolution (ICE), and the intraspecific Red Queen'.[2] The paper reads, in part: 'Perpetual ICE between the Y and the rest of the genome can thereby continually erode the genetic quality of the Y via genetic hitchhiking of mildly deleterious mutations. The decay of the Y is due to genetic hitchhiking, but it is the ICE process that acts in a catalytic way to continually drive the male versus female anatagonistic coevolution.' Even if most of this is Greek to you, there are certain words that catch the eye: words like 'enemies' and 'antagonism'. Then there is a recent textbook on the same material. Its title, quite simply, is '*Evolution: the four billion year war*'.[3] What is going on?

At some point in our past, our ancestors switched from the common reptilian habit of determining sex by the temperature of the egg to determining it genetically. The probable reason for the switch was so that each sex could start training for its special role at conception. In our case, the sex-determining gene made us male and the lack of it left us female, whereas in birds it happened the other way round. The gene soon attracted to its side other genes that benefited males: genes for big muscles, say, or aggressive tendencies. But because these were not wanted in females – wasting energy they would prefer to spend on offspring – these secondary genes found themselves at an advantage in one sex and at a disadvantage in the other. They are known in the trade as sexually antagonistic genes.

The dilemma was solved when another mutant gene suppressed the normal process of swapping of genetic material between the two paired chromosomes. Now the sexually antagonistic genes could diverge and go their different ways. The version on the Y chromosome could use calcium to make antlers; the version on the X chromosome could use calcium to make milk. Thus, a pair of middle-sized chromosomes, once home to all sorts of 'normal' genes, was hijacked by the process of sex determination and became the sex chromosomes, each attracting different sets of genes. On the Y chromosome, genes accumulate that benefit males but are often bad for females; on the X accumulate genes that are good for females and deleterious in males. For instance, there is a newly discovered gene called DAX, found on the X chromosome. A few rare people are born with one X and one Y chromosome, but with two copies of the DAX gene on the X chromosome. The result is, that although such people are genetically male, they develop into normal females. The reason, it transpires, is that DAX and SRY – the gene on the Y chromosome that makes men into men – are antagonistic to each other. One SRY defeats one DAX, but two DAXes defeat one SRY.[4]

This outbreak of antagonism between genes is a dangerous situation. Lurching into metaphor, one might begin to discern that the two chromosomes no longer have each other's interests at heart, let alone those of the species as a whole. Or, to put it more correctly, something can be good for the spread of a gene on the X chromosome that actually damages the Y chromosome or vice versa.

Suppose, for instance, that a gene appeared on the X chromosome that specified the recipe for a lethal poison that killed only sperm carrying Y chromosomes. A man with such a gene would have no fewer children than another man. But he would have all daughters and no sons. All of those daughters would carry the new gene, whereas if he had had sons as well, none of them would have carried it. Therefore, the gene is twice as common in the next generation as it would otherwise be. It would spread very rapidly. Such a gene would only cease to spread when it had exterminated so many males that the very survival of the species was in jeopardy and males were at a high premium.[5]

Far-fetched? Not at all. In the butterfly *Acrea encedon*, that is exactly what has happpened. The sex ratio is ninety-seven per cent female as a result. This is just one of many cases known of this form of evolutionary conflict, known as sex-chromosome drive. Most known instances are confined to insects, but only because scientists have looked more closely at insects. The strange language of conflict used in the remarks I quoted above now begins to make more sense. A piece of simple statistics: because females have two X chromosomes while males have an X and a Y, three-quarters of all sex chromosomes are Xs; one-quarter are Ys. Or, to put it another way, an X chromosome spends two-thirds of its time in females, and only one-third in males. Therefore, the X chromosome is three times as likely to evolve the ability to take pot shots at the Y as the Y is to evolve the ability to take pot shots at the X. Any gene on the Y chromosome is vulnerable to attack by a newly evolved driving X gene. The result has been that the Y chromosome has shed as many genes as possible and shut down the rest, to 'run away and hide' (in the technical jargon used by William Amos of Cambridge University).

So effectively has the human Y chromosome shut down most of its genes that the great bulk of its length consists of non-coding DNA, serving no purpose at all – but giving few targets for the X chromosome genes to aim at. There is a small region that seems to have slipped across from the X chromosome fairly recently, the so-called pseudo-autosomal region, and then there is one immensely important gene, the *SRY* gene mentioned above. This gene begins the whole cascade of events that leads to the masculinisation of the embryo. Rarely can a single gene have acquired such power. Although it only throws a switch, much else follows from that. The genitals grow to look like a penis and testes, the shape and constitution of the body are altered from female (the default in our species, though not in birds and butterflies), and various hormones go to work on the brain. There was a spoof map of the Y chromosome published in the journal *Science* a few years ago, which purported to have located genes for such stereotypically male traits as flipping between television channels, the ability to remember and tell jokes, an interest in the sports pages of newspapers, an addiction to death and destruction movies and an inability to express affection over the phone – among others. The joke is funny, though, only because we recognise these habits as male, and therefore far from mocking the idea that such habits are genetically determined, the joke reinforces the idea. The only thing wrong with the diagram is that these male behaviours come not from specific genes for each of them, but from the general masculinisation of the brain by hormones such as testosterone which results in a tendency to behave this way in the modern environment. Thus, in a sense, many masculine habits are all the products of the *SRY* gene itself, which sets in train the series of events that lead to the masculinisation of the brain as well as the body.

The *SRY* gene is peculiar. Its sequence is remarkably consistent between different men: there are virtually no point mutations (i.e., one-letter spelling differences) in the human race. *SRY* is, in that sense, a variation-free gene that has changed almost not at all since the last common ancestor of all people 200,000 years ago or so. Yet our *SRY* is very different from that of a chimpanzee, and different again from that of a gorilla: there is, between species, ten times as much variation in this gene as is typical for other genes. Compared with other active (i.e., expressed) genes, *SRY* is one of the fastest evolving.

How do we explain this paradox? According to William Amos and John Harwood, the answer lies in the process of fleeing and hiding that they call selective sweeps. From time to time, a driving gene appears on the X chromosome that attacks the Y chromosome by recognising the protein made by *SRY*. At once there is a selective advantage for any rare *SRY* mutant that is sufficiently different to be unrecognised. This mutant begins to spread at the expense of other males. The driving X chromosome distorts the sex ratio in favour of females but the spread of the new mutant *SRY* restores the balance. The end result is a brand new *SRY* gene sequence shared by all members of the species, with little variation. The effect of this sudden burst of evolution (which might happen so quickly as to leave few traces in the evolutionary record) would be to produce *SRY*s that were very different between species, but very similar within species. If Amos and Harwood are right, at least one such sweep must have occurred since the splitting of chimp ancestors and human ancestors, five to ten million years ago, but before the ancestor common to all modern human beings, 200,000 years ago.[6]

You may be feeling a little disappointed. The violence and conflict that I promised at the beginning of the chapter turn out to be little more than a detailed piece of molecular evolution. Fear not. I am not finished yet, and I plan to link these molecules to real, human conflict soon enough.

The leading scholar of sexual antagonism is William Rice of the University of California at Santa Cruz and he has completed a remarkable series of experiments to make the point explicit. Let us go back to our putative ancestral creature that has just acquired a distinct Y chromosome and is in the process of shutting down many of the genes on it to escape driving X genes. This nascent Y chromosome, in Rice's phrase, is now a hotspot for male-benefit genes. Because a Y chromosome will never find itself in a female, it is free to acquire genes that are very bad for females so long as they are at least slightly good for males (if you still thought evolution was about the good of the species, stop thinking so right now). In fruit flies, and for that matter in human beings, male ejaculate consists of sperm cells suspended in a rich soup called the seminal fluid. Seminal fluid contains proteins, products of genes. Their purpose is entirely unknown, but Rice has a shrewd idea. During fruit-fly sex, those proteins enter the bloodstream of the female and migrate to, among other places, her brain. There they have the effect of reducing the female's sexual appetite and increasing her ovulation rate. Thirty years ago, we would have explained that increase in terms of the good of the species. It is time for the female to stop seeking sexual partners and instead seek a nesting site. The male's seminal fluid redirects her behaviour to that end. You can hear the *National Geographic* commentary. Nowadays, this information takes on a more sinister aura. The male is trying to manipulate the female into mating with no other males and into laying more eggs for his sperm and he is doing so at the behest of sexually antagonistic genes, probably on the Y chromosome (or switched on by genes on the Y chromosome). The female is under selective pressure to be more and more resistant to such manipulation. The outcome is a stalemate.

Rice did an ingenious experiment to test his idea. For twenty-nine generations, he prevented female flies from evolving resistance: he kept a separate strain of females in which no evolutionary change occurred. Meanwhile, he allowed males to generate more and more effective seminal fluid proteins by testing them against more and more resistant females. After twenty-nine generations he brought the two lines together again. The result was a walkover. Male sperm was now so effective at manipulating female behaviour that it was effectively toxic: it could kill the females.[7]

Rice now believes that sexual antagonism is at work in all sorts of environments. It leaves its signature as rapidly evolving genes. In the shellfish the abalone, for instance, the lysin protein that the sperm uses to bore a hole through the glycoprotein matrix of the egg is encoded by a gene that changes very rapidly (the same is probably true in us), probably because there is an arms race between the lysin and the matrix. Rapid penetration is good for sperm but bad for the egg, because it allows parasites or second sperm through. Coming slightly closer to home, the placenta is controlled by rapidly evolving genes (and paternal ones, at that). Modern evolutionary theorists, led by David Haig, now think of the placenta as more like a parasitic takeover of the mother's body by paternal genes in the foetus. The placenta tries, against maternal resistance, to control her blood-sugar levels and blood pressure to the benefit of the foetus.[8] More on this in the chapter on chromosome 15.

But what about courtship behaviour? The traditional view of the peacock's elaborate tail is that it is a device designed to seduce females and that it is in effect designed by ancestral females' preferences. Rice's colleague, Brett Holland, has a different explanation. He thinks peacocks did indeed evolve long tails to seduce females, but that they did so because females grew more and more resistant to being so seduced. Males in effect use courtship displays as a substitute for physical coercion and females use discrimination to retain control over their own frequency and timing of mating. This explains a startling result from two species of wolf spiders. One species has tufts of bristles on its forelegs that it uses in courtship. Shown a video of a male spider displaying, the female will indicate by her behaviour whether the display turns her on. If the videos are altered so that the males' tufts disappear, the female is still just as likely to find the display arousing. But in another species, where there are no tufts, the artificial addition of tufts to males on the video more than doubled the acceptance rate of females. In other words, females gradually evolve so that they are turned off, not on, by the displays of males of their own species. Sexual selection is thus an expression of sexual antagonism between genes for seduction and genes for resistance.[9]

Rice and Holland come to the disturbing conclusion that the more social and communicative a species is, the more likely it is to suffer from sexually antagonistic genes, because communication between the sexes provides the medium in which sexually antagonistic genes thrive. The most social and communicative species on the planet is humankind. Suddenly it begins to make sense why relations between the human sexes are such a minefield, and why men have such vastly different interpretations of what constitutes sexual harassment from women. Sexual relations are driven not by what is good, in evolutionary terms, for men or for women, but for their chromosomes. The ability to seduce a woman was good for Y chromosomes in the past; the ability to resist seduction by a man was good for X chromosomes in the past.

This kind of conflict between complexes of genes (the Y chromosome being one such complex), does not just apply to sex. Suppose that there is a version of a gene that increases the telling of lies (not a very realistic proposition, but there might be a large set of genes that affect truthfulness indirectly). Such a gene might thrive by making its possessors into successful con-artists. But then suppose there is also a version of a different gene (or set of genes) that improves the detecting of lies, perhaps on a different chromosome. That gene would thrive to the extent that it enabled its possessors to avoid being taken in by con-artists. The two would evolve antagonistically, each gene encouraging the other, even though it would be quite possible for the same person to possess both. There is between them what Rice and Holland call 'interlocus contest evolution', or ICE. Exactly such a competitive process probably did indeed drive the growth of human intelligence over the past three million years. The notion that our brains grew big to help us make tools or start fires on the savannah has long since lost favour. Instead, most evolutionists believe in the Machiavellian theory – that bigger brains were needed in an arms race between manipulation and resistance to manipulation. 'The phenomena we refer to as intelligence may be a byproduct of intergenomic conflict between genes mediating offense and defense in the context of language', write Rice and Holland.[10]

Forgive the digression into intelligence. Let's get back to sex. Probably one of the most sensational, controversial and hotly disputed genetic discoveries was the announcement by Dean Hamer in 1993 that he had found a gene on the X chromosome that had a powerful influence on sexual orientation, or, as the media quickly called it, 'a gay gene'.[11] Hamer's study was one of several published about same time all pointing towards the conclusion that homosexuality was 'biological' – as opposed to being the consequence of cultural pressure or conscious choice. Some of this work was done by gay men themselves, such as the neuroscientist Simon LeVay of the Salk Institute, keen to establish in the public mind what they were convinced about in their own minds: that homosexuals were 'born that way'. They believed, with some justice, that prejudice would be less against a lifestyle that was not a deliberate 'choice' but an innate propensity. A genetic cause would also make homosexuality seem less threatening to parents by making it clear that gay role models could not turn youths gay unless they had the propensity already. Indeed conservative intolerance of homosexuality has recently taken to attacking the evidence for its genetic nature. 'We should be careful about accepting the claim that some are "born to be gay", not just because it is untrue, but because it provides leverage to homosexual rights organisations', wrote the Conservative Lady Young in the *Daily Telegraph* on 29 July 1998.

But however much some of the researchers may have desired a particular outcome, the studies are objective and sound. There is no room for doubt that homosexuality is highly heritable. In one study, for example, among fifty-four gay men who were fraternal twins, there were twelve whose twin was also gay; and among fifty-six gay men who were identical twins, there were twenty-nine whose twin was also gay. Since twins share the same environment, whether they are fraternal or identical, such a result implies that a gene or genes accounts for about half of the tendency for a man to be gay. A dozen other studies came to a similar conclusion.[12]

Intrigued, Dean Hamer decided to seek the genes that were involved. He and his colleagues interviewed 110 families with gay male members and noticed something unusual. Homosexuality seemed to run in the female line. If a man was gay, the most likely other member of the previous generation to be gay was not his father but his mother's brother.

That immediately suggested to Hamer that the gene might be on the X chromosome, the only set of nuclear genes a man inherits exclusively from his mother. By comparing a set of genetic markers between gay men and straight men in the families in his sample, he quickly found a candidate region in Xq28, the tip of the long arm of the chromosome. Gay men shared the same version of this marker seventy-five per cent of the time; straight men shared a different version of the marker seventy-five per cent of the time. Statistically, that ruled out coincidence with ninety-nine per cent confidence. Subsequent results reinforced the effect, and ruled out any connection between the same region and lesbian orientation.[13]

To canny evolutionary biologists, such as Robert Trivers, the suggestion that such a gene might lie on the X chromosome immediately rang a bell. The problem with a gene for sexual orientation is that the version that causes homosexuality would quite quickly become extinct. Yet it is plainly present in the modern population at a significant level. Perhaps four per cent of men are definitively gay (and a smaller percentage bisexual). Since gay men, are, on average, less likely to have children than straight men, the gene would be doomed to have long since dwindled in frequency to vanishing point unless it carried some compensating advantage.

Trivers argued that, because an X chromosome spends twice as much time in women as it does in men, a sexually antagonistic gene that benefited female fertility could survive even if it had twice as large a deleterious effect on male fertility. Suppose, for example, that the gene Hamer had found determined age of puberty in women, or even something like breast size (remember, this is just a thought experiment). Each of those characteristics might affect female fertility. Back in the Middle Ages, large breasts might mean more milk, or might attract a richer husband whose children were less likely to die in infancy. Even if the same version of the same gene reduced male fertility by making sons attracted to other men, such a gene could survive because of the advantage it gave daughters.

Until Hamer's gene itself is found and decoded, the link between homosexuality and sexual antagonism is no more than a wild guess. Indeed, it remains a possibility that the connection between Xq28 and sexuality is misleading. Michael Bailey's recent research on homosexual pedigrees has failed to find a maternal bias to be a general feature. Other scientists, too, have failed to find Hamer's link with Xq28. At present it looks as if it may have been confined to those families Hamer studied. Hamer himself cautions that until the gene is in the bag, it is a mistake to assume otherwise.[14]

Besides, there is now a complicating factor: a completely different explanation of homosexuality. It is becoming increasingly clear that sexual orientation correlates with birth order. A man with one or more elder brothers is more likely to be gay than a man with no siblings, only younger siblings, or with one or more elder sisters. The birth order effect is so strong that each additional elder brother increases the probability of homosexuality by roughly one-third (this can still mean a low probability: an increase from three to four per cent is an increase of thirty-three per cent). The effect has now been reported in Britain, the Netherlands, Canada and the United States, and in many different samples of people.[15]

For most people, the first thought would be a quasi-Freudian one: that something in the dynamics of growing up in a family with elder brothers might predispose you towards homosexuality. But, as so often, the Freudian reaction is almost certainly the wrong one. (The old Freudian idea that homosexuality was caused by a protective mother and a distant father almost certainly confused cause and effect: the boy's developing effeminate interests repel the father and the mother becomes overprotective in compensation.) The answer probably lies, once more, in the realm of sexual antagonism.

An important clue lies in the fact that there is no such birth-order effect for lesbians, who are randomly distributed within their families. In addition, the number of elder sisters is also irrelevant in predicting male homosexuality. There is something specific to occupying a womb that has already held other males which increases the probability of homosexuality. The best explanation concerns a set of three active genes on the Y chromosome called the H-Y minor histocompatibility antigens. One of these genes encodes a protein called anti-Mullerian hormone, a substance vital to the masculinisation of the body: it causes the regression of the Mullerian ducts in the male embryo – these being the precursors of the womb and Fallopian tubes. What the other two genes do is not certain. They are not essential for the masculinisation of the genitals, which is achieved by testosterone and anti-Mullerian hormone alone. The significance of this is now beginning to emerge.

The reason these gene products are called antigens is because they are known to provoke a reaction from the immune system of the mother. As a result, the immune reaction is likely to be stronger in successive male pregnancies (female babies do not produce H-Y antigens, so do not raise the immune reaction). Ray Blanchard, one of those who studies the birth-order effect, argues that the H-Y antigens' job is to switch on other genes in certain tissues, in particular in the brain – and indeed there is good evidence that this is true in mice. If so, the effect of a strong immune reaction against these proteins from the mother would be partly to prevent the masculinisation of the brain, but not that of the genitals. That in turn might cause them to be attracted to other males, or at least not attracted to females. In an experiment in which baby mice were immunised

against H-Y antigens, they grew up to be largely incapable of successful mating, compared with controls, though frustratingly the
experimenter did not report the reasons why. Likewise, male fruit
flies can be irreversibly induced to show only female sexual
behaviour by the switching on at a crucial point in development of
a gene called 'transformer'.[16]

People are not mice or flies, and there is plenty of evidence that
the sexual differentiation of the human brain continues after birth.
Homosexual men are clearly not, except in very rare cases, 'mental'
women trapped inside 'physical' men. Their brains must have been
at least partly masculinised by hormones. It remains possible, however, that they missed some hormone during some early and crucial
sensitive period and that this permanently affects some functions,
including sexual orientation.

The man who first set in train the ideas that led to sexual antagonism, Bill Hamilton, understood how profoundly it shook our notions
of what genes are: 'There had come the realisation', he wrote later,
'that the genome wasn't the monolithic data bank plus executive
team devoted to one project – keeping oneself alive, having babies
– that I had hitherto imagined it to be. Instead, it was beginning to
seem more a company boardroom, a theatre for a power struggle
of egoists and factions.' Hamilton's new understanding of his genes
began to affect his understanding of his mind:[17]

My own conscious and seemingly indivisible self was turning out far from
what I had imagined and I need not be so ashamed of my self-pity! I was
an ambassador ordered abroad by some fragile coalition, a bearer of
conflicting orders from the uneasy masters of a divided empire . . . As I
write these words, even so as to be able to write them, I am pretending
to a unity that, deep inside myself, I now know does not exist. I am
fundamentally mixed, male with female, parent with offspring, warring
segments of chromosomes that interlocked in strife millions of years
before the River Severn ever saw the Celts and Saxons of Housman's
poem ['A Shropshire Lad'].

The idea of genes in conflict with each other, the notion of the genome as a sort of battlefield between parental genes and childhood genes, or between male genes and female genes, is a little-known story outside a small group of evolutionary biologists. Yet it has profoundly shaken the philosophical foundations of biology.

Self-Interest

We are survival machines – robot vehicles blindly programmed to preserve the selfish molecules known as
genes. This is a truth that still fills me with astonishment.
Richard Dawkins, The Selfish Gene

Instruction manuals that come with new gadgets are notoriously
frustrating. They never seem to have the one piece of information
you need, they send you round in circles, they leave you high and
dry, and they definitely lose something in the translation from
Chinese. But at least they do not insert, just when you are getting
to the bit that matters, five copies of Schiller's 'Ode to Joy' or a
garbled version of a set of instructions for how to saddle a horse.
Nor do they (generally) include five copies of a complete set of
instructions for how to build a machine that would copy out just
that set of instructions. Nor do they break the actual instructions
you seek into twenty-seven different paragraphs interspersed with
long pages of irrelevant junk so that even finding the right instructions is a massive task. Yet that is a description of the human

retinoblastoma gene and, as far as we know, it is typical of human genes: twenty-seven brief paragraphs of sense interrupted by twenty-six long pages of something else.

Mother Nature concealed a dirty little secret in the genome. Each gene is far more complicated than it needs to be, it is broken up into many different 'paragraphs' (called exons) and in between lie long stretches (called introns) of random nonsense and repetitive bursts of wholly irrelevant sense, some of which contain real genes of a completely different (and sinister) kind.

The reason for this textual confusion is that the genome is a book that wrote itself, continually adding, deleting and amending over four billion years. Documents that write themselves have unusual properties. In particular, they are prone to parasitism. Analogies become far-fetched at this point, but try to imagine a writer of instruction manuals who arrives at his computer each morning to find paragraphs of his text clamouring for his attention. The ones that shout loudest bully him into including another five copies of themselves on the next page he writes. The true instructions still have to be there, or the machine will never be assembled, but the manual is full of greedy, parasitic paragraphs taking advantage of the writer's compliance.

Actually, with the advent of email, the analogy is no longer as far-fetched as it once was. Suppose I sent you an email that read: 'Beware, there is a nasty computer virus about; if you open a message with the word "marmalade" in the title, it will erase your hard disk! Please pass this warning on to everybody you can think of.' I made up the bit about the virus; there are, so far as I know, no emails called 'marmalade' doing the rounds. But I have very effectively hijacked your morning and caused you to send on my warning. My email was the virus.[1]

So far, each chapter of this book has concentrated on a gene or genes, tacitly assuming that they are the things that matter in the genome. Genes, remember, are stretches of DNA that comprise the recipe for proteins. But ninety-seven per cent of our genome does not consist of true genes at all. It consists of a menagerie of strange entities called pseudogenes, retropseudogenes, satellites, minisatellites, microsatellites, transposons and retrotransposons: all collectively known as 'junk DNA', or sometimes, probably more accurately, as 'selfish DNA'. Some of these are genes of a special kind, but most are just chunks of DNA that are never transcribed into the language of protein. Since the story of this stuff follows naturally from the tale of sexual conflict related in the last chapter, this chapter will be devoted to junk DNA.

Fortunately this is a good place to tell the story, because I have nothing more particular to say about chromosome 8. That is not to imply that it is a boring chromosome, or that it possesses few genes, just that none of the genes yet found on chromosome 8 has caught my rather impatient attention. (For its size, chromosome 8 has been relatively neglected, and is one of the least mapped chromosomes.) Junk DNA is found on every chromosome. Yet, ironically, junk DNA is the first part of the human genome that has found a real, practical, everyday use in the human world. It has led to DNA fingerprinting.

Genes are protein recipes. But not all protein recipes are desirable. The commonest protein recipe in the entire human genome is the gene for a protein called reverse transcriptase. Reverse transcriptase is a gene that serves no purpose at all as far as the human body is concerned. If every copy of it were carefully and magically removed from the genome of a person at the moment of conception, the person's health, longevity and happiness would be more likely to be improved than damaged. Reverse transcriptase is vital for a certain kind of parasite. It is an extremely useful – nay essential – part of the genome of the AIDS virus: a crucial contributor to its ability to infect and kill its victims. For human beings, in contrast, the gene is a nuisance and a threat. Yet it is one of the commonest genes in the whole genome. There are several hundred copies of it, possibly thousands, spread about the human chromosomes. This is an astonishing fact, akin to discovering that the commonest use of cars is for getting away from crimes. Why is it there?

A clue comes from what reverse transcriptase does. It takes an

RNA copy of a gene, copies it back into DNA and stitches it back into the genome. It is a return ticket for a copy of a gene. By this means the AIDS virus can integrate a copy of its own genome into human DNA the better to conceal it, maintain it and get it efficiently copied. A good many of the copies of the reverse transcriptase gene in the human genome are there because recognisable 'retroviruses' put them there, long ago or even relatively recently. There are several thousand nearly complete viral genomes integrated into the human genome, most of them now inert or missing a crucial gene. These 'human endogenous retroviruses' or Hervs, account for 1.3% of the entire genome. That may not sound like much, but 'proper' genes account for only 3%. If you think being descended from apes is bad for your self-esteem, then get used to the idea that you are also descended from viruses.

But why not cut out the middle man? A viral genome could drop most of the virus's genes and keep just the reverse transcriptase gene. Then this streamlined parasite could give up the laborious business of trying to jump from person to person in spit or during sex, and instead just hitchhike down the generations within its victims' genomes. A true genetic parasite. Such 'retrotransposons' are far commoner even than retroviruses. The commonest of all is a sequence of 'letters' known as a LINE-1. This is a 'paragraph' of DNA, between a thousand and six thousand 'letters' long, that includes a complete recipe for reverse transcriptase near the middle. LINE-1s are not only very common – there may be 100,000 copies of them in each copy of your genome – but they are also gregarious, so that the paragraph may be repeated several times in succession on the chromosome. They account for a staggering 14.6% of the entire genome, that is, they are nearly five times as common as 'proper' genes. The implications of this are terrifying. LINE-1s have their own return tickets. A single LINE-1 can get itself transcribed, make its own reverse transcriptase, use that reverse transcriptase to make a DNA copy of itself and insert that copy anywhere among the genes. This is presumably how there come to be so many copies of LINE-1 in the first place. In other words, this repetitive 'paragraph' of 'text' is there because it is good at getting itself duplicated – no other reason.

'A flea hath smaller fleas that on him prey; and these have smaller fleas to bite 'em, and so proceed ad infinitum.' If LINE-1s are about, they, too, can be parasitised by sequences that drop the reverse transcriptase gene and use the ones in LINE-1s. Even commoner than LINE-1s are shorter 'paragraphs' called Alus. Each Alu contains between 180 and 280 'letters', and seems to be especially good at using other people's reverse transcriptase to get itself duplicated. The Alu text may be repeated a million times in the human genome – amounting to perhaps ten per cent of the entire 'book'.[2]

For reasons that are not entirely clear, the typical Alu sequence bears a close resemblance to a real gene, the gene for a part of a protein-making machine called the ribosome. This gene, unusually, has what is called an internal promoter, meaning that the message 'READ ME' is written in a sequence in the middle of the gene. It is thus an ideal candidate for proliferation, because it carries the signal for its own transcription and does not rely on landing near another such promoter sequence. As a result, each Alu gene is probably a 'pseudogene'. Pseudogenes are, to follow a common analogy, rusting wrecks of genes that have been holed below the waterline by a serious mutation and sunk. They now lie on the bottom of the genomic ocean, gradually growing rustier (that is, accumulating more mutations) until they no longer even resemble the gene they once were. For example, there is a rather nondescript gene on chromosome 9, which, if you take a copy of it and then probe the genome for sequences that resemble this gene, you will find at fourteen locations on eleven chromosomes: fourteen ghostly hulks that have sunk. They were redundant copies that, one after another, mutated and stopped being used. The same may well be true of most genes – that for every working gene, there are a handful of wrecked copies elsewhere in the genome. The interesting thing about this particular set of fourteen is that they have been sought not just in people, but in monkeys, too. Three of the human pseudogenes were sunk after the split between Old-World monkeys and

New-World monkeys. That means, say the scientists breathlessly, they were relieved of their coding functions 'only' around thirty-five million years ago.[3]

Alus have proliferated wildly, but they too have done so in comparatively recent times. Alus are found only in primates, and are divided into five different families, some of which have appeared only since the chimpanzees and we parted company (that is, within the last five million years). Other animals have different short repetitive 'paragraphs'; mice have ones called B1s.

All this information about LINE-1s and Alus amounts to a major and unexpected discovery. The genome is littered, one might almost say clogged, with the equivalent of computer viruses, selfish, parasitic stretches of letters which exist for the pure and simple reason that they are good at getting themselves duplicated. We are full of digital chain letters and warnings about marmalade. Approximately thirty-five per cent of human DNA consists of various forms of selfish DNA, which means that replicating our genes takes thirty-five per cent more energy than it need. Our genomes badly need worming.

Nobody suspected this. Nobody predicted that when we read the code for life we would find it so riddled with barely controlled examples of selfish exploitation. Yet we should have predicted it, because every other level of life is parasitised. There are worms in animals' guts, bacteria in their blood, viruses in their cells. Why not retrotransposons in their genes? Moreover, by the mid-1970s, it was dawning on many evolutionary biologists, especially those interested in behaviour, that evolution by natural selection was not much about competition between species, not much about competition between groups, not even mostly about competition between individuals, but was about competition between genes using individuals and occasionally societies as their temporary vehicles. For instance, given the choice between a safe, comfortable and long life for the individual or a risky, tiring and dangerous attempt to breed, virtually all animals (and indeed plants) choose the latter. They choose to shorten their odds of death in order to have offspring. Indeed, their bodies are designed with planned obsolescence called ageing that causes them to decay after they reach breeding age – or, in the case of squid or Pacific salmon, to die at once. None of this makes any sense unless you view the body as a vehicle for the genes, as a tool used by genes in their competition to perpetuate themselves. The body's survival is secondary to the goal of getting another generation started. If genes are 'selfish replicators' and bodies are their disposable 'vehicles' (in Richard Dawkins's controversial terminology), then it should not be much of a surprise to find some genes that achieve their replication without building their own bodies. Nor should it be a surprise to find that genomes, like bodies, are habitats replete with their own version of ecological competition and co-operation. Truly, in the 1970s for the first time, evolution became genetic.

To explain the fact that the genome contained huge gene-less regions, two pairs of scientists suggested in 1980 that these regions were replete with selfish sequences whose only function was survival within the genome. 'The search for other explanations may prove', they said, 'if not intellectually sterile, ultimately futile.' For making this bold forecast, they were much mocked at the time. Geneticists were still stuck in the mindset that if something were in the human genome it must serve a human purpose, not a selfish purpose of its own. Genes were just protein recipes. It made no sense to think of them as having goals or dreams. But the suggestion has been spectacularly vindicated. Genes do indeed behave as if they have selfish goals, not consciously, but retrospectively: genes that behave in this way thrive and genes that don't don't.[4]

A segment of selfish DNA is not just a passenger, whose presence adds to the size of the genome and therefore to the energy cost of copying the genome. Such a segment is also a threat to the integrity of genes. Because selfish DNA is in the habit of jumping from one location to another, or sending copies to new locations, it is apt to land in the middle of working genes, messing them up beyond recognition, and then jumping out again causing the mutation to revert. This was how transposons were first discovered, in the late 1940s, by the far-sighted and much neglected geneticist Barbara McClintock (she was eventually awarded the Nobel prize in 1983).

She noticed that mutations in the colour of maize seeds occur in such a manner that can only be explained by mutations jumping into and out of pigment genes.[5]

In human beings, LINE-1s and Alus have caused mutations by landing in the middle of all sorts of genes. They have caused haemophilia, for instance, by landing in clotting-factor genes. But, for reasons that are not well understood, as a species we are less troubled by DNA parasites than some other species. Approximately 1 in every 700 human mutations is caused by 'jumping genes', whereas in mice nearly ten per cent of mutations are caused by jumping genes. The potential danger posed by jumping genes was dramatically illustrated by a sort of natural experiment in the 1950s in the tiny fruit fly, *Drosophila*. The fruit fly is the favourite experimental animal for geneticists. The species they study, called *Drosophila melanogaster*, has been transported all over the world to be bred in laboratories. It has frequently escaped and has met other, native species of fruit fly. One of these species, called *Drosophila willistoni*, carries a jumping gene called a P element. Somehow in about 1950, somewhere in South America, perhaps via a blood-sucking mite, *Drosophila willistoni*'s jumping gene entered the *Drosophila melanogaster* species. (One of the great concerns attached to so-called 'xeno-transplants' of organs from pigs or baboons is that they might unleash a new form of jumping gene upon our species, like the P element of fruit flies.) The P element has since spread like wildfire, so that most fruit flies have the P element, though not those collected from the wild before 1950 and kept in isolation since. The P element is a piece of selfish DNA that shows its presence by disrupting the genes into which it jumps. Gradually, the rest of the genes in the fruit fly's genome have fought back, inventing ways of suppressing the P element's jumping habit. The P elements are settling down as passengers.

Human beings possess nothing so sinister as a P element, at least not at the moment. But a similar element, called 'sleeping beauty', has been found in salmon. Introduced into human cells in the laboratory it thrives, demonstrating cut-and-paste ability. And something similar to the spread of the P element probably happened with each of the nine human Alu elements. Each spread through the species, disrupting genes until the other genes asserted their common interest in suppressing it, whereupon it settled down in its present fairly quiescent state. What we see in the human genome is not some rapidly advancing parasitic infection, but the dormant cysts of many past parasites, each of which spread rapidly until the genome found a way of suppressing them, but not excising them.

In this respect (as in others) we seem to be more fortunate than fruit flies. We appear to have a general mechanism for suppressing selfish DNA, at least if you believe a controversial new theory. The suppression mechanism goes by the name of cytosine methylation. Cytosine is the letter C of the genetic code. Methylating it (literally by attaching a methyl group of carbon and hydrogen atoms) prevents it from being transcribed by the reader. Much of the genome spends large chunks of the time in the methylated – blocked – state, or rather most gene promoters do (the parts at the beginning of the gene where transcription starts). It has generally been assumed that methylation serves to switch off genes that are not needed in particular tissues, thus making the brain different from the liver, which is different from the skin and so on. But a rival explanation is gaining ground. Methylation may have almost nothing to do with tissue-specific expression and much to do with suppressing transposons and other intragenomic parasites. Most methylation lies within transposons such as Alu and LINE-1. The new theory holds that during the early development of the embryo, all genes are briefly stripped of any methylation and switched on. This is then followed by a close inspection of the whole genome by molecules whose job is to spot repetitive sequences and close them down with methylation. In cancer tumours, one of the first things to happen is demethylation of the genes. As a result, the selfish DNA is released from its handcuffs and richly expressed in tumours. Since they are good at messing up other genes, these transposons then make the cancer worse. Methylation, according to this argument, serves to suppress the effect of selfish DNA.[6]

LINE-1 is generally about 1,400 'letters' long. Alu is generally at least 180 'letters' long. There are, however, sequences even shorter than Alu that also accumulate in vast, repetitive stutters. It is perhaps too far-fetched to call these shorter sequences parasites, but they proliferate in roughly the same manner – that is, they are there because they contain a sequence that is good at getting itself duplicated. It is one of these short sequences that has a practical use in forensic and other sciences. Meet the 'hypervariable minisatellite'. This neat little sequence is found on all the chromosomes; it crops up at more than one thousand locations in the genome. In every case the sequence consists of a single 'phrase', usually about twenty 'letters' long, repeated over and over again many times. The 'word' can vary according to the location and the individual, but it usually contains the same central 'letters': GGGCAGGAXG (where X can be any 'letter'). The significance of this sequence is that it is very similar to one that is used by bacteria to initiate the swapping of genes with other bacteria of the same species, and it seems to be involved in the encouragement of gene swapping between chromosomes in us as well. It is as if each sequence is a sentence with the words 'SWAP ME ABOUT' in the middle.

Here is an example of a repetition of a minisatellite: hxckswapmeaboutlopl-hxckswapmeaboutlopl-hxckswapmeaboutlopl-hxckswapmeaboutlopl-hxckswapmeaboutlopl-hxckswapmeaboutlopl-hxckswapmeaboutlopl-hxckswapmeaboutlopl-hxckswapmeaboutlopl-hxckswapmeaboutlopl. Ten repeats in this case. Elsewhere, at each of one thousand locations, there might be fifty or five repeats of the same phrase. Following instructions, the cell starts swapping the phrases with the equivalent series on the other copy of the same chromosome. But in doing so it makes fairly frequent mistakes, adding or subtracting to the number of repeats. In this way each series of repeats gradually changes length, fast enough so that it is different in every individual, but slowly enough so that people mostly have the same repeat lengths as their parents. Since there are thousands of series, the result is a unique set of numbers for each individual.

Alec Jeffreys and his technician Vicky Wilson stumbled on mini-satellites in 1984, largely by accident. They were studying how genes evolve by comparing the human gene for the muscle protein myoglobin with its equivalent from seals when they noticed a stretch of repetitious DNA in the middle of the gene. Because each minisatellite shares the same core sequence of twelve letters, but because the number of repeats can vary so much, it is a relatively simple matter to fish out this minisatellite array and compare the size of the array in different individuals. It turns out that the repeat number is so variable that everybody has a unique genetic fingerprint: a string of black marks looking just like a bar code. Jeffreys immediately spotted the significance of what he had found. Neglecting the myoglobin genes that were the target of his study, he started investigating what could be done with unique genetic fingerprints. Because strangers have such different genetic fingerprints, immigration authorities were immediately interested in testing the claims of would-be immigrants that they were close relatives of people already in the country. Genetic fingerprinting proved that they were generally telling the truth, which eased much misery. But a more dramatic use was to follow soon after.[7]

On 2 August 1986, a young schoolgirl's body was found in a thorn thicket close to the village of Narborough, in Leicestershire. Dawn Ashworth, aged fifteen, had been raped and strangled. A week later, the police arrested a young hospital porter, Richard Buckland, who confessed to the murder. There the matter would have rested. Buckland would have gone to prison, convicted of the killing. However, the police were anxious to clear up an unsolved case, of a girl named Lynda Mann, also fifteen, also from Narborough, also raped, strangled and left in an open field, but nearly three years before. The murders were so similar it seemed implausible that they had not been committed by the same man. But Buckland refused to confess to Mann's murder.

Word of Alec Jeffreys's fingerprinting breakthrough had reached the police via the newspapers, and since he worked in Leicester, less than ten miles from Narborough, the local police contacted

Jeffreys and asked him if he could confirm the guilt of Buckland in the Mann case. He agreed to try. The police supplied him with semen taken from both girls' bodies and a sample of Buckland's blood.

Jeffreys had little difficulty finding various minisatellites in each sample. After more than a week's work the genetic fingerprints were ready. The two semen samples were identical and must have come from the same man. Case closed. But what Jeffreys saw next astonished him. The blood sample had a radically different fingerprint from the semen samples: Buckland was not the murderer.

The Leicestershire police protested heatedly that this was an absurd conclusion and that Jeffreys must have got it wrong. Jeffreys repeated the test and so did the Home Office forensic laboratory, with exactly the same result. Reluctantly, the baffled police withdrew the case against Buckland. For the first time in history a man was exonerated on the basis of his DNA sequences.

But nagging doubts remained. Buckland had, after all, confessed and policemen would find genetic fingerprinting a lot more convincing if it could convict the guilty as well as acquit the innocent. So, five months after Ashworth's death, the police set out to test the blood of 5,500 men in the Narborough area to look for a genetic fingerprint that matched that of the murdering rapist's sperm. No sample matched.

Then a man who worked in a Leicester bakery named Ian Kelly happened to remark to his colleagues that he had taken the blood test even though he lived nowhere near Narborough. He had been asked to do so by another worker in the bakery, Colin Pitchfork, who did live in Narborough. Pitchfork claimed to Kelly that the police were trying to frame him. One of Kelly's colleagues repeated the tale to the police, who arrested Pitchfork. Pitchfork quickly confessed to killing both girls, but this time the confession proved true: the DNA fingerprint of his blood matched that of the semen found on both bodies. He was sentenced on 23 January 1988 to life in prison.

Genetic fingerprinting immediately became one of forensic

science's most reliable and potent weapons. The Pitchfork case, an extraordinary virtuoso demonstration of the technique, set the tone for years to come: genetic fingerprinting's ability to acquit the innocent, even in the face of what might seem overwhelming evidence of guilt; its ability to flush out the guilty just by the threat of its use; its amazing precision and reliability – if properly used; its reliance on small samples of bodily tissue, even nasal mucus, spit, hair or bone from a long-dead corpse.

Genetic fingerprinting has come a long way in the decade since the Pitchfork case. In Britain alone, by mid-1998 320,000 samples of DNA had been collected by the Forensic Science Service and used to link 28,000 people to crime scenes. Nearly twice as many samples have been used to exonerate innocent people. The technique has been simplified, so that single sites of minisatellites can be used instead of many. Genetic fingerprinting has also been amplified, so that tiny minisatellites or even microsatellites can be used to give unique 'bar codes'. Not only the lengths but the actual sequences of the minisatellite repeats can be analysed to give greater sophistication. Such DNA typing has also been misused or discredited in court, as one might expect when lawyers are involved. (Much of the misuse reflects public naïvety with statistics, rather than anything to do with the DNA: nearly four times as many potential jurors will convict if told that a DNA match has a chance probability of 0.1 per cent than if told one in a thousand men match the DNA – yet they are the same facts.[8])

DNA fingerprinting has revolutionised not just forensic science but all sorts of other fields as well. It was used to confirm the identity of the exhumed corpse of Josef Mengele in 1990. It was used to confirm the presidential parenthood of the semen on Monica Lewinsky's dress. It was used to identify the illegitimate descendants of Thomas Jefferson. It has so blossomed in the field of paternity testing, both by officials publicly and by parents privately, that in 1998 a company called Identigene placed billboards by freeways all over America reading: 'WHO'S THE FATHER? CALL 1-800-DNA-TYPE'. They received 300 calls a day asking for their

$600 tests, both from single mothers trying to demand child-support from the 'fathers' of their children and from suspicious 'fathers' unsure if their partner's children were all theirs. In more than two-thirds of cases the DNA evidence showed that the mother was telling the truth. It is a moot point whether the offence caused to some fathers by discovering that their partners were unfaithful outweighs the reassurance others receive that their suspicions were unfounded. Britain, predictably, had a fierce media row when the first such private service set up shop: in Britain such medical technologies are supposed to remain the property of the state, not the individual.[9]

More romantically, the application of genetic fingerprinting to paternity testing has revolutionised our understanding of bird song. Have you ever noticed that thrushes, robins and warblers continue singing long after they have paired up in spring? This flies in the face of the conventional notion that bird song's principal function is the attraction of a mate. Biologists began DNA-testing birds in the late 1980s, trying to determine which male had fathered which chicks in each nest. They discovered, to their surprise, that in the most monogamous of birds, where just one male and one female faithfully help each other to rear the brood, the female mates quite often with neighbouring males other than their ostensible 'spouses'. Cuckoldry and infidelity are much, much commoner than anybody expected (because they are committed in great secrecy). DNA fingerprinting led to an explosion of research into a richly rewarding theory known as sperm competition, which can explain such trivia as the fact that chimpanzee testicles are four times the size of gorilla testicles, even though chimpanzees are one-quarter the size of gorillas. Male gorillas monopolise their mates, so their sperm meets no competitors; male chimpanzees share their mates, so each needs to produce large quantities of sperm and mate frequently to increase his chances of being the father. It also explains why male birds sing so hard when already 'married'. They are looking for 'affairs'.[10]

Disease

A desperate disease requires a dangerous remedy.

Guy Fawkes

On chromosome 9 lies a very well-known gene: the gene that determines your ABO blood group. Since long before there was DNA fingerprinting, blood groups have appeared in court. Occasionally, the police get lucky and match the blood of the criminal to blood found at the scene of the crime. Blood grouping presumes innocence. That is to say, a negative result can prove you were not the murderer absolutely, but a positive one can only suggest that you might be the murderer.

Not that this logic had much impact on the California Supreme Court, which in 1946 ruled that Charlie Chaplin was most definitely the father of a certain child despite unambiguous proof from the incompatibility of their blood groups that he could not have been. But then judges were never very good at science. In paternity suits as well as murder cases, blood grouping, like genetic fingerprinting, or indeed fingerprinting, is the friend of the innocent. In the days of DNA fingerprinting, blood-group forensics is redundant. Blood groups are much more important in transfusion, though again in a wholly negative way: receiving the wrong blood can be fatal. And blood groups can give us insights into the history of human migrations, though once more they have been almost entirely superseded in this role by other genes. So you might think blood groups are rather dull. You would be wrong. Since 1990 they have found an entirely new role: they promise understanding of how and why our genes are all so different. They hold the key to human polymorphism.

The first and best known of the blood group systems is the ABO system. First discovered in 1900, this system originally had three different names with confusing consequences: type I blood, according to Moss's nomenclature was the same as type IV blood according to Jansky's nomenclature. Sanity gradually prevailed and the nomenclature adopted by the Viennese discoverer of the blood groups became universal: A, B, AB and O. Karl Landsteiner expressively described the disaster that befell a wrong transfusion thus: 'lytischen und agglutinierenden Wirkungen des Blutserums'. The red cells all stick together. But the relation between the blood groups was not simple. People with type A blood could safely donate to those with A or AB; those with B could donate to those with B and AB; those with AB could donate only to those with AB; and those with O blood could donate to anybody – O is therefore known as the universal donor. Nor was there any obvious geographic or racial reason underlying the different types. Roughly forty per cent of Europeans have type O blood, forty per cent have type A blood, fifteen per cent have type B blood and five per cent have type AB blood. The proportions are similar in other continents, with the marked exception of the Americas, where the native American population was almost exclusively type O, save for some Canadian tribes, who were very often type A, and Eskimos, who were sometimes type AB or B.

It was not until the 1920s that the genetics of the ABO blood groups fell into place, and not until 1990 that the gene involved came to light. A and B are 'co-dominant' versions of the same gene, O being the 'recessive' form of it. The gene lies on chromosome

9, near the end of the long arm. Its text is 1,062 'letters' long, divided into six short and one long exons ('paragraphs') scattered over several 'pages' – 18,000 letters in all – of the chromosome. It is a medium-sized gene, then, interrupted by five longish introns. The gene is the recipe for galactosyl transferase,[1] an enzyme, i.e. a protein with the ability to catalyse a chemical reaction.

The difference between the A gene and the B gene is seven letters out of 1,062, of which three are synonymous or silent: that is, they make no difference to the amino acid chosen in the protein chain. The four that matter are letters 523, 700, 793 and 800. In people with type A blood these letters read C, G, C, G. In people with type B blood they read G, A, A, C. There are other, rare differences. A few people have some of the A letters and some of the B letters, and a rare version of the A type exists in which a letter is missing near the end. But these four little differences are sufficient to make the protein sufficiently different to cause an immune reaction to the wrong blood.[2]

The O group has just a single spelling change compared with A, but instead of a substitution of one letter for another, it is a deletion. In people with type O blood, the 258th letter, which should read 'G', is missing altogether. The effect of this is far-reaching, because it causes what is known as a reading-shift or frame-shift mutation, which is far more consequential. (Recall that if Francis Crick's ingenious comma-free code of 1957 had been correct, reading-shift mutations would not have existed.) The genetic code is read in three-letter words and has no punctuation. An English sentence written in three-letter words might read something like: the fat cat sat top mat and big dog ran bit cat. Not exactly poetry, I admit, but it will do. Change one letter and it still makes fairly good sense: the fat xat sat top mat and big dog ran bit cat. But delete the same letter instead, and read the remaining letters in groups of three, and you render the whole sentence meaningless: the fat ats att opm ata ndb igd ogr anb itc at. This is what has happened to the ABO gene in people with the O blood group. Because they lack just one letter fairly early in the message, the whole subsequent message says something completely different. A different protein is made with different properties. The chemical reaction is not catalysed.

This sounds drastic, but it appears to make no difference at all. People with type O blood are not noticeably disadvantaged in any walk of life. They are not more likely to get cancer, be bad at sports, have little musical ability or something. In the heyday of eugenics, no politician called for the sterilisation of people with the O blood group. Indeed, the remarkable thing about blood groups, the thing that has made them so useful and so politically neutral, is that they seem to be completely invisible; they correlate with nothing.

But this is where things get interesting. If blood groups are invisible and neutral, then how did they evolve to the present state? Was it pure chance that landed the inhabitants of the Americas with type O blood? At first glance the blood groups seem to be an example of the neutral theory of evolution, promulgated by Motoo Kimura in 1968: the notion that most genetic diversity is there because it makes no difference, not because it has been picked by natural selection for a purpose. Kimura's theory was that mutation pumps a continual stream of mutations that do not affect anything into the gene pool, and that they are gradually purged again by genetic drift – random change. So there is constant turnover without adaptive significance. Return to earth in a million years and large chunks of the human genome would read differently for entirely neutral reasons.

'Neutralists' and 'selectionists' for a while grew quite exercised about their respective beliefs, and when the dust settled Kimura was left with a respectable following. Much variation does indeed seem to be neutral in its effects. In particular, the closer scientists look at how proteins change, the more they conclude that most changes do not affect the 'active site' where the protein does its chemical tricks. In one protein, there have been 250 genetic changes since the Cambrian age between one group of creatures and another, yet only six of them matter at all.[3]

But we now know the blood groups are not as neutral as they seem. There is indeed a reason behind them. From the early 1960s, it gradually became apparent that there was a connection between blood groups and diarrhoea. Children with type A blood fell victim to certain strains of infant diarrhoea but not to others; children with type B blood fell victim to other strains; and so on. In the late 1980s, people with the O group were discovered to be much more susceptible to infection with cholera. Dozens of studies later, the details grow more distinct. Not only are those people with type O blood susceptible, but those with A, B and AB differ in their susceptibility. The most resistant people are those with the AB genotype, followed by A, followed by B. All of these are much more resistant than those with O. So powerful is this resistance in AB people that they are virtually immune to cholera. It would be irresponsible to say that people with type AB blood can safely drink from a Calcutta sewer – they might get another disease – but it is true that even if these people did pick up the *Vibrio* bacterium that causes cholera and it settled in their gut, they would not get diarrhoea.

Nobody yet knows how the AB genotype offers protection against this most virulent and lethal of human diseases, but it presents natural selection with an immediate and fascinating problem. Remember that we each have two copies of each chromosome, so A people are actually AAs, that is they have an A gene on each of their ninth chromosomes, and B people are actually BBs. Now imagine a population with just these three kinds of blood groups: AA, BB and AB. The A gene is better for cholera resistance than the B gene. AA people are therefore likely to have more surviving children than BB people. Therefore the B gene is likely to die out – that's natural selection. But it doesn't happen like that, because AB people survive best of all. So the healthiest children will be the offspring of AAs and BBs. All their children will be AB, the most cholera-resistant type. But even if an AB mates with another AB, only half their children will be AB; the rest will be AA and BB, the latter being the most susceptible type. It is a world of strangely fluctuating fortunes. The very combination that is most beneficial in your generation guarantees you some susceptible children.

Now imagine what happens if everybody in one town is AA, but a newcomer arrives who is BB. If she can fend off the cholera long enough to breed, she will have AB children, who will be resistant. In other words, the advantage will always lie with the rare version of the gene, so neither version can become extinct because if it becomes rare, it comes back into fashion. This is known, in the trade, as frequency-dependent selection, and it seems to be one of the commonest reasons that we are all so genetically diverse.

This explains the balance between A and B. But if O blood makes you more susceptible to cholera, then why has natural selection not driven the O mutation extinct? The answer probably lies with a different disease, malaria. People with type O blood seem to be slightly more resistant to malaria than people of other blood groups. They also seem to be slightly less likely to get cancers of various kinds. This enhanced survival was probably enough to keep the O version of the gene from disappearing, despite its association with susceptibility to cholera. A rough balance was struck between the three variations on the blood group gene.

The link between disease and mutations was first noticed in the late 1940s by an Oxford graduate student with a Kenyan background, Anthony Allison. He suspected that the frequency of a disease called sickle-cell anaemia in Africa might be connected with the prevalence of malaria. The sickle-cell mutation, which causes blood cells to collapse in the absence of oxygen, is frequently fatal to those with two copies of it, but only mildly harmful to those with just one copy. But those with one copy are largely resistant to malaria. Allison tested the blood of Africans living in malarial areas and found that those with the mutation were far less likely to have the malaria parasite as well. The sickle-cell mutation is especially common in parts of west Africa where malaria has long been endemic, and is common also in African-Americans, some of whose ancestors came from west Africa in the slave ships. Sickle-cell disease is a high price paid today for malaria resistance in the past. Other forms of anaemia, such as the thalassaemia common in various parts of the Mediterranean and south-east Asia, appear to have a similar protective effect against malaria, accounting for its presence in regions once infested with the disease.

The haemoglobin gene, where the sickle-cell mutation occurs as just a single-letter change, is not alone in this respect. According to one scientist, it is the tip of an iceberg of genetic resistance to malaria. Up to twelve different genes may vary in their ability to confer resistance to malaria. Nor is malaria alone. At least two genes vary in their ability to confer resistance to tuberculosis, including the gene for the vitamin D receptor, which is also associated with a variability in susceptibility to osteoporosis. 'Naturally', writes Adrian Hill of Oxford University,[4] 'We can't resist suggesting that natural selection for TB resistance in the recent past may have increased the prevalence of susceptibility genes for osteoporosis.'

Meanwhile, a newly discovered but similar connection links the genetic disease cystic fibrosis with the infectious disease typhoid. The version of the CFTR gene on chromosome 7 that causes cystic fibrosis – a dangerous disease of the lungs and intestines – protects the body against typhoid, an intestinal disease caused by a *Salmonella* bacterium. People with just one such version do not get cystic fibrosis, but they are almost immune to the debilitating dysentery and fever caused by typhoid. Typhoid needs the usual version of the CFTR gene to get into the cells it infects; the altered version, missing three DNA letters, is no good to it. By killing those with other versions of the gene, typhoid put natural pressure on the altered version to spread. But because people inheriting two copies of the altered version were lucky to survive at all, the gene could never be very common. Once again, a rare and nasty version of a gene was maintained by disease.[5]

Approximately one in five people are genetically unable to release the water-soluble form of the ABO blood group proteins into their saliva and other body fluids. These 'non-secretors' are more likely to suffer from various forms of disease, including meningitis, yeast infection and recurrent urinary tract infection. But they are less likely to suffer from influenza or respiratory syncitial virus. Wherever you look, the reasons behind genetic variability seem to have something to do with infectious disease.[6]

We have barely scratched the surface of this subject. As they scourged our ancestors, the great epidemic diseases of the past – plague, measles, smallpox, typhus, influenza, syphilis, typhoid, chicken pox, and others – left behind their imprint on our genes. Mutations which granted resistance thrived, but that resistance often came at a price, the price varying from severe (sickle-cell anaemia) to theoretical (the inability to receive transfusions of the wrong type of blood).

Indeed, until recently, doctors were in the habit of underestimating the importance of infectious disease. Many diseases that are generally thought to be due to environmental conditions, occupation, diet or pure chance are now beginning to be recognised as the side-effects of chronic infections with little known viruses or bacteria. The most spectacular case is stomach ulcers. Several drug companies grew rich on new drugs intended to fight the symptoms of ulcers, when all that were needed all along were antibiotics. Ulcers are caused by *Helicobacter pylori*, a bacterium usually acquired in childhood, rather than by rich food, anxiety or misfortune. Likewise, there are strong suggestive links between heart disease and infection with chlamydia or herpes virus, between various forms of arthritis and various viruses, even between depression or schizophrenia and a rare brain virus called Borna disease virus that usually infects horses and cats. Some of these correlations may prove misleading and in other cases the disease may attract the microbe rather than the other way round. But it is a proven fact that people vary in their genetic resistance to things like heart disease. Perhaps these genetic variants, too, relate to resistance to infection.[7]

In a sense the genome is a written record of our pathological past, a medical scripture for each people and race. The prevalence of O blood groups in native Americans may reflect the fact that cholera and other forms of diarrhoea, which are diseases associated with crowded and insanitary conditions, never established themselves in the newly populated continents of the western hemisphere before relatively modern times. But then cholera was a rare disease probably confined to the Ganges delta before the 1830s, when it suddenly spread to Europe, the Americas and Africa. We need a better explanation of the puzzling prevalence of the O version of the gene in native Americans, especially given the fact that the blood of ancient pre-Columbian mummies from North America seems quite often to be of the A or B type. It is almost as if the A and B genes were rapidly driven extinct by a different selection pressure unique to the western hemisphere. There are hints that the cause might be syphilis, a disease that seems to be indigenous to the Americas (this is still hotly disputed in medical-history circles, but the fact remains that syphilitic lesions are known in North American skeletons from before 1492, but not in European skeletons from before that date). People with the O version of the gene seem to be less susceptible to syphilis than those with other blood types.[8]

Now consider a bizarre discovery that would have made little sense before the discovery of the association between susceptibility to cholera and blood groups. If, as a professor, you ask four men and two women each to wear a cotton T-shirt, no deodorant and no perfume, for two nights, then hand these T-shirts to you, you will probably be humoured as a mite kinky. If you then ask a total of 121 men and women to sniff the armpits of these dirty T-shirts and rank them according to attractiveness of smell, you will be considered, to put it mildly, eccentric. But true scientists should not be embarrassable. The result of exactly such an experiment, by Claus Wederkind and Sandra Füri, was the discovery that men and women most prefer (or least dislike) the body odour of members of the opposite sex who are most different from them genetically. Wederkind and Füri looked at MHC genes on chromosome 6, which are the genes involved in the definition of self and the recognition of parasitic intruders by the immune system. They are immensely variable genes. Other things being equal, a female mouse will prefer to mate with a male that has maximally different MHC genes from herself, a fact she discerns by sniffing his urine. It was this discovery that alerted Wederkind and Füri to the possibility that we, too, might retain some such ability to choose our mates on the basis of their genes. Only women on the contraceptive pill failed to show a clear preference for different MHC genotypes in male-impregnated T-shirt armpits. But then the pill is known to affect the sense of smell. As Wedekind and Füri put it,[9] 'No one smells good to everybody; it depends on who is sniffing whom.'

The mouse experiment had always been interpreted in terms of outbreeding: the female mouse tries to find a male from a genetically different population, so that she can have offspring with varied genes and little risk of inbred diseases. But perhaps she – and T-shirt-sniffing people – are actually doing something that makes sense in terms of the blood-group story. Remember that, when making love in a time of cholera, an AA person is best off looking for a BB mate, so that all their children will be cholera-resistant ABs. If the same sort of system applies to other genes and their co-evolution with other diseases – and the MHC complex of genes seems to be the principal site of disease-resistance genes – then the advantage of being sexually attracted to a genetic opposite is obvious.

The Human Genome Project is founded upon a fallacy. There is no such thing as 'the human genome'. Neither in space nor in time can such a definite object be defined. At hundreds of different loci, scattered throughout the twenty-three chromosomes, there are genes that differ from person to person. Nobody can say that the blood group A is 'normal' and O, B and AB are 'abnormal'. So when the Human Genome Project publishes the sequence of the typical human being, what will it publish for the ABO gene on chromosome 9? The project's declared aim is to publish the average or 'consensus' sequence of 200 different people. But this would miss the point in the case of the ABO gene, because it is a crucial part of its function that it should not be the same in everybody. Variation is an inherent and integral part of the human – or indeed any – genome.

Nor does it make sense to take a snapshot at this particular moment in 1999 and believe that the resulting picture somehow represents a stable and permanent image. Genomes change.

Different versions of genes rise and fall in popularity driven often by the rise and fall of diseases. There is a regrettable human tendency to exaggerate stability, to believe in equilibrium. In fact the genome is a dynamic, changing scene. There was a time when ecologists believed in 'climax' vegetation – oak forests for England, fir forests for Norway. They have learnt better. Ecology, like genetics, is not about equilibrium states. It is about change, change and change. Nothing stays the same forever.

The first person who half glimpsed this was probably J. B. S. Haldane, who tried to find a reason for the abundance of human genetic variation. As early as 1949 he conjectured that genetic variation might owe a good deal to the pressures of parasites. But Haldane's Indian colleague, Suresh Jayakar, rocked the boat in 1970 by suggesting that there need be no stability, and that parasites could cause a perpetual cycling fluctuation in gene frequencies. By the 1980s the torch had passed to the Australian Robert May, who demonstrated that even in the simplest system of a parasite and its host, there might be no equilibrium outcome: that eternal chaotic motion could flow from a deterministic system. May thus became one of the fathers of chaos theory. The baton was picked up by the Briton William Hamilton, who developed mathematical models to explain the evolution of sexual reproduction, models that relied upon a genetic arms race between parasites and their hosts, and which resulted in what Hamilton called 'the permanent unrest of many [genes]'.[10]

Some time in the 1970s, as happened in physics half a century before, the old world of certainty, stability and determinism in biology fell. In its place we must build a world of fluctuation, change and unpredictability. The genome that we decipher in this generation is but a snapshot of an ever-changing document. There is no definitive edition.

CHROMOSOME 10

Stress

This is the excellent foppery of the world, that, when we are sick in fortune – often the surfeit of our own behaviour, – we make guilty of our disasters the sun, the moon, and the stars; as if we were villains by necessity, fools by heavenly compulsion . . . an admirable evasion of whoremaster man, to lay his goatish disposition to the charge of a star.   *William Shakespeare*, King Lear

The genome is a scripture in which is written the past history of plagues. The long struggles of our ancestors with malaria and dysentery are recorded in the patterns of human genetic variation. Your chances of avoiding death from malaria are pre-programmed in your genes, and in the genes of the malaria organism. You send out your team of genes to play the match, and so does the malaria parasite. If their attackers are better than your defenders, they win. Bad luck. No substitutes allowed.

But it is not like that, is it? Genetic resistance to disease is the last resort. There are all sorts of simpler ways of defeating disease. Sleep under a mosquito net, drain the swamps, take a pill, spray

DDT around the village. Eat well, sleep well, avoid stress, keep your immune system in good health and generally maintain a sunny disposition. All of these things are relevant to whether you catch an infection. The genome is not the only battlefield. In the last few chapters I have fallen into the habit of reductionism. I have taken the organism apart to isolate its genes and discern their particular interests. But no gene is an island. Each one exists as part of an enormous confederation called the body. It is time to put the organism back together again. It is time to visit a much more social gene, a gene whose whole function is to integrate some of the many different functions of the body, and a gene whose existence gives the lie to the mind–body dualism that plagues our mental image of the human person. The brain, the body and the genome are locked, all three, in a dance. The genome is as much under the control of the other two as they are controlled by it. That is partly why genetic determinism is such a myth. The switching on and off of human genes can be influenced by conscious or unconscious external action.

Cholesterol – a word pregnant with danger. The cause of heart disease; bad stuff; red meat. You eat it, you die. Nothing could be more wrong than this equation of cholesterol with poison. Cholesterol is an essential ingredient of the body. It lies at the centre of an intricate system of biochemistry and genetics that integrates the whole body. Cholesterol is a small organic compound that is soluble in fat but not in water. The body manufactures most of its cholesterol from sugars in the diet, and could not survive without it. From cholesterol at least five crucial hormones are made, each with a very different task: progesterone, aldosterone, cortisol, testosterone and oestradiol. Collectively, they are known as the steroids. The relationship between these hormones and the genes of the body is intimate, fascinating and unsettling.

Steroids have been used by living creatures for so long that they probably pre-date the split between plants, animals and fungi. The hormone that triggers the shedding of an insect's skin is a steroid. So is the enigmatic chemical known in human medicine as vitamin D. Some synthetic, or anabolic, steroids can be manufactured to trick the body into suppressing inflammation, while others can be used for building athletes' muscles. Yet other steroids, derived originally from plants, can mimic human hormones sufficiently well to be used as oral contraceptives. Others still, products of the chemical industry, may be responsible for the feminisation of male fish in polluted streams and the falling sperm counts of modern men.

There is a gene on chromosome 10 called *CYP17*. It makes an enzyme, which enables the body to convert cholesterol into cortisol, testosterone and oestradiol. Without the enzyme, the pathway is blocked and the only hormones that can be made from cholesterol are progesterone and corticosterone. People who lack a working copy of this gene cannot make other sex hormones so they fail to go through puberty; if genetically male, they look like girls.

But put the sex hormones on one side for a moment and consider the other hormone that is made using *CYP17*: cortisol. Cortisol is used in virtually every system in the body, a hormone that literally integrates the body and the mind by altering the configuration of the brain. Cortisol interferes with the immune system, changes the sensitivity of the ears, nose and eyes, and alters various bodily functions. When you have a lot of cortisol coursing through your veins, you are – by definition – under stress. Cortisol and stress are virtually synonymous.

Stress is caused by the outside world, by an impending exam, a recent bereavement, something frightening in the newspaper or the unremitting exhaustion of caring for a person with Alzheimer's disease. Short-term stressors cause an immediate increase in epinephrine and norepinephrine, the hormones that make the heart beat faster, the feet go cold. These hormones prepare the body for 'fight or flight' in an emergency. Stressors that last for longer activate a different pathway that results in a much slower, but more persistent increase in cortisol. One of cortisol's most surprising effects is that it suppresses the working of the immune system. It is a remarkable fact that people who have been preparing for an important exam, and have shown the symptoms of stress, are more likely to catch colds and other infections, because one of the effects of cortisol is to reduce the activity, number and lifetime of lymphocytes – white blood cells.

Cortisol does this by switching genes on. It only switches on genes in cells that have cortisol receptors in them, which have in turn been switched on by some other triggers. The genes that it switches on mostly switch on other genes in turn, and sometimes the genes that they switch on will then switch on other genes and so on. The secondary effects of cortisol can involve tens, or maybe even hundreds, of genes. But the cortisol was only made in the first place because a series of genes was switched on in the adrenal cortex to make the enzymes necessary for making cortisol – among them *CYP17*. It is a system of mindboggling complexity: if I started to list even the barest outlines of the actual pathways I would bore you to tears. Suffice to say that you cannot produce, regulate and respond to cortisol without hundreds of genes, nearly all of which work by switching each other on and off. It is a timely lesson that the main purpose of most genes in the human genome is regulating the expression of other genes in the genome.

I promised not to bore you, but let me just take a quick glimpse at one of the effects of cortisol. In white blood cells cortisol is almost certainly involved in switching on a gene called *TCF*, also on chromosome 10, thus enabling *TCF* to make its own protein, whose job is to suppress the expression of another protein called interleukin 2, and interleukin 2 is a chemical that puts white blood cells on alert to be especially vigilant for germs. So cortisol suppresses the immune alertness of white blood cells and makes you more susceptible to disease.

The question I want to put in front of you is: who's in charge? Who ordered all these switches to be set in the right way in the first place, and who decides when to start to let loose the cortisol? You could argue that the genes are in charge, because the differentiation of the body into different cell types, each with different genes switched on, was at root a genetic process. But that's misleading, because genes are not the cause of stress. The death of a loved one, or an impending exam do not speak directly to the genes. They are information processed by the brain.

So the brain is in charge. The hypothalamus of the brain sends out the signal that tells the pituitary gland to release a hormone that tells the adrenal gland to make and secrete cortisol. The hypothalamus takes its orders from the conscious part of the brain which gets its information from the outside world.

But that's not much of an answer either, because the brain is part of the body. The reason the hypothalamus stimulates the pituitary which stimulates the adrenal cortex is not because the brain decided or learnt that this was a good way to do things. It did not set up the system in such a way that thinking about an impending exam would make you less resistant to catching a cold. Natural selection did that (for reasons I will come back to shortly). And in any case, it is a wholly involuntary and unconscious reaction, which implies that it is the exam, rather than the brain, that is in charge of events. And if the exam is in charge, then society is to blame, but what is society but a collection of individuals, which brings us back to bodies? Besides, people vary in their susceptibility to stress. Some find impending exams terrifying, others take them in their stride. What is the difference? Somewhere down the cascade of events that is the production, control and reaction to cortisol, stress-prone people must have subtly different genes from phlegmatic folk. But who or what controls these genetic differences?

The truth is that nobody is in charge. It is the hardest thing for human beings to get used to, but the world is full of intricate, cleverly designed and interconnected systems that do not have control centres. The economy is such a system. The illusion that economies run better if somebody is put in charge of them – and decides what gets manufactured where and by whom – has done devastating harm to the wealth and health of peoples all over the world, not just in the former Soviet Union, but in the west as well. From the Roman Empire to the European Union's high-definition television initiative, centralised decisions about what to invest in have been disastrously worse than the decentralised chaos of the market. Economies are not centralised systems; they are markets with decentralised, diffuse controls.

It is the same with the body. You are not a brain running a body by switching on hormones. Nor are you a body running a genome by switching on hormone receptors. Nor are you a genome running a brain by switching on genes that switch on hormones. You are all of these at once.

Many of the oldest arguments in psychology boil down to misconceptions of this kind. The arguments for and against 'genetic determinism' presuppose that the involvement of the genome places it above and beyond the body. But as we have seen it is the body that switches on genes when it needs them, often in response to a more or less cerebral, or even conscious, reaction to external events. You can raise your cortisol levels just by thinking about stressful eventualities – even fictional ones. Likewise, the dispute between those who believe that a certain suffering is purely psychiatric and those who insist it has a physical cause – consider ME, or chronic fatigue syndrome – is missing the point entirely. The brain and the body are part of the same system. If the brain, responding to psychological stress, stimulates the release of cortisol and cortisol suppresses the reactivity of the immune system, then a dormant viral infection may well flare up, or a new one catch hold. The symptoms may indeed be physical and the causes psychological. If a disease affects the brain and alters the mood, the causes may be physical and the symptoms psychological.

This topic is known as psychoneuroimmunology, and it is slowly inching its way into fashion, mostly resisted by doctors and mostly hyped by faith healers of one kind or another. But the evidence is real enough. Chronically unhappy nurses have more episodes of cold sores than others who also carry the virus. People with anxious personalities have more outbreaks of genital herpes than sunny optimists. At West Point military academy, the students most likely to catch mononucleosis (glandular fever), and the ones most likely to get a severe illness from it if they do, are the ones who are most anxious and pressured by their work. Those who care for Alzheimer's patients (an especially stressful activity) have fewer disease-fighting T lymphocytes in their blood than expected. Those who lived near Three Mile Island nuclear plant at the time of its accident had more cancers than expected three years later, not because they were exposed to radiation (they weren't), but because their cortisol levels had risen, reducing the responsiveness of their immune system to cancer cells. Those bereaved by the death of a spouse have a less responsive immune system for several weeks afterwards. Children whose families have been riven by a parental argument in the previous week are more likely to catch viral infections. People with most psychological stress in their past get more colds than people who have led happy lives. And if you find these sorts of studies hard to believe, then most of them have been replicated in some form or another using mice or rats.[1]

Poor old René Descartes usually gets the blame for the dualism that has dominated western thinking and made us all so resistant to the idea that the mind can affect the body and the body can affect the mind, too. He barely deserves the blame for an error we all commit. In any case, the fault is not so much dualism – the notion of a separate mind detached from the material matter of the brain. There is a far greater fallacy that we all commit, so easily that we never even notice it. We instinctively assume that bodily biochemistry is cause whereas behaviour is effect, an assumption we have taken to a ridiculous extent in considering the impact of genes upon our lives. If genes are involved in behaviour then it is they that are the cause and they that are deemed immutable. This is a mistake made not just by genetic determinists, but by their vociferous opponents, the people who say behaviour is 'not in the genes'; the people who deplore the fatalism and predestination implied, they say, by behaviour genetics. They give too much ground to their opponents by allowing this assumption to stand, for they tacitly admit that if genes are involved at all, then they are at the top of the hierarchy. They forget that genes need to be switched on, and external events – or free-willed behaviour – can switch on genes. Far from us lying at the mercy of our omnipotent genes, it is often our genes that lie at the mercy of us. If you go bungee jumping or take a stressful job, or repeatedly imagine a terrible fear, you will raise your cortisol levels, and the cortisol will dash about the body busy switching on genes. (It is also an indisputable fact that you can trigger activity in the 'happiness centres' of the brain with a deliberate smile, as surely as you trigger a smile with happy thoughts. It really does make you feel better to smile. The physical can be at the beck and call of the behavioural.)

Some of the best insights into the way behaviour alters gene expression come from studies of monkeys. Fortunately for those who believe in evolution, natural selection is an almost ridiculously thrifty designer and once she has hit upon a system of genes and hormones to indicate and respond to stress, she is loath to change it (we are ninety-eight per cent chimpanzees and ninety-four per cent baboons, remember). So the very same hormones work in the very same way in monkeys and switch on the very same genes. There is a troop of baboons in east Africa whose bloodstream cortisol levels have been closely studied. When a certain young male baboon attached himself to a new troop, as male baboons of a certain age are wont to do, he became highly aggressive as he fought to establish himself in the hierarchy of his chosen society. The result was a steep increase in the cortisol concentration in his blood as well as that of his unwilling hosts. As his cortisol (and testosterone) levels rose, so his lymphocyte count fell. His immune system bore the brunt of his behaviour. At the same time his blood began to contain less and less of the cholesterol bound to high-density lipoprotein (HDL). Such a fall is a classic precursor of furring up of the coronary arteries. Not only was the baboon, by his free-willed behaviour, altering his hormones, and hence the expression of his genes, he was thereby increasing his risk of both infection and coronary artery disease.[2]

Among monkeys kept in zoos, the ones whose arteries fur up are the ones at the foot of the pecking order. Bullied by their more senior colleagues, they are continuously stressed, their blood is rich in cortisol, their brains are low in serotonin, their immune systems are permanently depressed and scar tissue builds up on the walls of their coronary arteries. Quite why is still a mystery. Many scientists now believe that coronary disease is at least partly caused by infectious agents, such as chlamydia bacteria and herpes viruses. The effect of stress is to lower immune surveillance of these dormant infections which allows them to flourish. Perhaps, in this sense, heart disease in monkeys is infectious, though stress may play a role as well.

People are very like monkeys. The discovery that monkeys low in the hierarchy get heart disease came soon after the far more startling discovery that British civil servants working in Whitehall also get heart disease in proportion to their lowliness in the bureaucratic pecking order. In a massive, long-term study of 17,000 civil servants, an almost unbelievable conclusion emerged: the status of a person's job was more able to predict their likelihood of a heart attack than obesity, smoking or high blood pressure. Somebody in a low-grade job, such as a janitor, was nearly four times as likely to have a heart attack as a permanent secretary at the top of the heap. Indeed, even if the permanent secretary was fat, hypertensive or a smoker, he was still less likely to suffer a heart attack at a given age than a thin, non-smoking, low-blood-pressure janitor. Exactly the same result emerged from a similar study of a million employees of the Bell Telephone Company in the 1960s.[3]

Think about this conclusion for a moment. It undermines almost everything you have ever been told about heart disease. It relegates cholesterol to the margins of the story (high cholesterol is a risk factor, but only in those with genetic predispositions to high cholesterol, and even in these people the beneficial effects of eating less fat are small). It relegates diet, smoking and blood pressure – all the physiological causes so preferred by the medical profession – to secondary causes. It relegates to a footnote the old and largely discredited notion that stress and heart failure come with busy, senior jobs or fast-living personalities: again there is a grain of truth in this fact, but not much. Instead, dwarfing these effects, science now elevates something non-physiological, something strictly related to the outside world: the status of your job. Your heart is at the mercy of your pay grade. What on earth is going on?

The monkeys hold the clue. The lower they are in the pecking order, the less control they have over their lives. Likewise in the civil service, cortisol levels rise in response not to the amount of work you do, but to the degree to which you are ordered about by other people. Indeed, you can demonstrate this effect experimentally, just by giving two groups of people the same task to do, but ordering one group to do the task in a set manner and to an imposed schedule. This externally controlled group of people suffers a greater increase in stress hormones and rise in blood pressure and heart rate than the other group.

Twenty years after the Whitehall study began, it was repeated in a department of the civil service that then began to experience privatisation. At the beginning of the study, the civil servants had no notion of what it meant to lose their jobs. Indeed, when a questionnaire was being piloted for the study, the subjects objected to a question that asked if they feared losing their jobs. It was a meaningless question in the civil service, they explained: at worst they might be transferred to a different department. By 1995 they knew exactly what losing their jobs meant; more than one in three had already experienced it. The effect of privatisation was to give everybody a feeling that their lives were at the mercy of external factors. Not surprisingly, stress followed and with stress came ill health – far more ill health than could be explained by any changes in diet, smoking or drinking.

The fact that heart disease is a symptom of lack of control explains a good deal about its sporadic appearance. It explains why so many people in senior jobs have heart attacks soon after they retire and 'take it easy'. From running offices they often move to lowly and menial jobs (washing dishes, walking the dog) in domestic environments run by their spouses. It explains why people are capable of postponing an illness, even a heart attack, until after a family wedding or a major celebration – until the end of a period of busy work when they are in control of events. (Students also tend to go down with illnesses *after* periods of acute exam pressure, not during them.) It explains why unemployment and welfare dependency are so good at making people ill. No alpha-male monkey was ever such an intransigent and implacable controller of subordinates' lives as the social services of the state are of people dependent on welfare. It may even explain why modern buildings in which the windows cannot be opened make people sicker than older buildings in which people have more control over their environment.

I am going to repeat myself for emphasis. Far from behaviour being at the mercy of our biology, our biology is often at the mercy of our behaviour.

What is true of cortisol is also true of other steroid hormones. Testosterone levels correlate with aggression, but is that because the hormone causes aggression, or because release of the hormone is caused by aggression? In our materialism, we find the first alternative far easier to believe. But in fact, as studies of baboons demonstrate, the second is closer to the truth. The psychological precedes the physical. The mind drives the body, which drives the genome.[4]

Testosterone is just as good at suppressing the immune system as cortisol. This explains why, in many species, males catch more diseases and have higher mortality than females. This immune suppression applies not just to the body's resistance to micro-organisms, but to large parasites, too. The warble fly lays its eggs on the skin of deer and cattle; the maggot then burrows into the flesh of the animal before returning to the skin to form a nodule in which to metamorphose into a fly. Reindeer in northern Norway are especially troubled by these parasites, but males noticeably more than females. On average, by the age of two, a male reindeer has three times as many warble-fly nodules in its skin as a female reindeer, yet castrated males have the same number as females. A similar pattern can be found for many infectious parasites, including, for instance, the protozoan that causes Chagas' disease, the affliction widely believed to explain Charles Darwin's chronic illnesses. Darwin was bitten by the bug that carries Chagas' disease while travelling in Chile and some of his later symptoms fit the disease. If Darwin had been a woman, he might have spent less time feeling sorry for himself.[5]

Yet it is to Darwin that we must turn for enlightenment here. The fact that testosterone suppresses immune function has been seized upon by a cousin of natural selection known as sexual selection and ingeniously exploited. In Darwin's second book on evolution, *The descent of man*, he put forward the notion that, just as a pigeon breeder can breed pigeons, so a female can breed males. By consistently choosing which males to mate with over many generations, female animals can alter the shape, colour, size or song of males of their species. Indeed, as I described in the chapter on chromosomes X and Y, Darwin suggested that this is exactly what has happened in the case of peacocks. It was not until a century later, in the 1970s and 1980s, that a series of theoretical and experimental studies demonstrated that Darwin was right, and that the tails, plumes, antlers, songs and size of male animals are bred into them by consistent trends of passive or active female choice, generation after generation.

But why? What conceivable benefit can a female derive from picking a male with a long tail or a loud song? Two favourite ideas have dominated the debate, the first being that the female must follow the prevailing fashion lest she have sons that are not themselves attractive to females who follow the prevailing fashion. The second idea, and the one that I propose to consider here, is that the quality of the male's ornament reflects the quality of his genes in some way. In particular, it reflects the quality of his resistance to prevailing infections. He is saying to all who would listen: see how strong I am; I can grow a great tail or sing a great song, because I am not debilitated by malaria, nor infected with worms. And the fact that testosterone suppresses the immune system is actually the greatest possible help in making this an honest message. For the quality of his ornaments depends on the level of testosterone in his blood: the more testosterone he has, the more colourful, large, songful or aggressive he will be. If he can grow a great tail despite lowering his immune defences, yet not catch disease, he must be impressive genetically. It is almost as if the immune system obscures the genes; testosterone parts the veil and allows the female to see directly into the genes.[6]

This theory is known as the immunocompetence handicap and

it depends upon the immune-suppressive effects of testosterone being unavoidable. A male cannot get round the handicap by raising his testosterone levels and not suppressing his immune system. If such a male existed, he would surely be a great success and would leave many offspring behind, because he could grow a long tail with (literally) immunity. Hence, the theory implies that the link between steroids and immune suppression is as fixed, inevitable and important as any in biology.

But this is even more puzzling. Nobody has a good explanation for the link in the first place, let alone its inevitability. Why should bodies be designed so that their immune systems are depressed by steroid hormones? It means that whenever you are stressed by a life event, you become more vulnerable to infection, cancer and heart disease. That is kicking you when you are down. It means that whenever an animal raises its testosterone level to fight its rivals for mates or to enhance its display, it becomes more vulnerable to infection, cancer and heart disease. Why?

Various scientists have struggled with this conundrum, but to little effect. Paul Martin, in his book on psychoneuroimmunology called *The sickening mind*, discusses two possible explanations and rejects them both. First is the notion that it is all a mistake, and that the links between the immune system and the stress response are accidental by-products of the way some other systems have to be designed. As Martin points out, this is a deeply unsatisfactory explanation for a system full of complex neural and chemical links. Very, very few parts of the body are accidental, vestigial or functionless, especially not complex parts. Natural selection would ruthlessly cull links that suppress the immune response if they had no function.

The second explanation, that modern life produces prolonged and unnatural stresses and that in an ancient environment such stresses would have been much shorter-lived, is equally disappointing. Baboons and peacocks live in a state of nature, yet they too – and virtually every other bird and mammal on the planet – suffer from immune suppression by steroids.

Martin admits to bafflement. He cannot explain the fact that stress inevitably depresses the immune system. Nor can I. Perhaps, as Michael Davies has suggested, the depression is designed to save energy in times of semi-starvation, a common form of stress before the modern era. Or perhaps the response to cortisol is a side-effect of the response to testosterone (they are very similar chemicals) and the response to testosterone is deliberately engineered into males by the genes of females the better to sort the fitter – that is more disease resistant – males from the less fit. In other words, the link may be the product of a kind of sexual antagonism like the one discussed in the chapter on chromosomes X and Y. I don't find this explanation convincing, so I challenge you to find a better one.

Personality

A man's character is his fate.

Heraclitus

The tension between universal characteristics of the human race and particular features of individuals is what the genome is all about. Somehow the genome is responsible for both the things we share with other people and the things we experience uniquely in ourselves. We all experience stress; we all experience the elevated cortisol that goes with it; we all suffer from the immune-suppressive effects thereof. We all have genes switched on and off by external events in this way. But each of us is unique, too. Some people are phlegmatic, some highly strung. Some are anxious, others risk-seeking. Some are confident, others shy. Some are quiet, others loquacious. We call these differences personality, a word that means more than just character. It means the innate and individual element in character.

To seek out the genes that influence personality, it is time to move from the hormones of the body to the chemicals of the mind – though the distinction is by no means a hard-and-fast one. On the short arm of chromosome 11, there lies a gene called *D4DR*.

It is the recipe for a protein called a dopamine receptor, and it is switched on in cells of certain parts of the brain but not in others. Its job is to stick out of the membrane of a neuron at the junction with another neuron (known as a synapse), ready to latch on to a small chemical called dopamine. Dopamine is a neurotransmitter, released from the tips of other neurons by an electrical signal. When the dopamine receptor encounters dopamine, it causes its own neuron to discharge an electrical signal of its own. That is the way the brain works: electrical signals that cause chemical signals that cause electrical signals. By using at least fifty different chemical signals, the brain can carry on many different conversations at once: each neurotransmitter stimulates a different set of cells or alters their sensitivity to different chemical messengers. It is misleading to think of a brain as a computer for many reasons, but one of the most obvious is that an electrical switch in a computer is just an electrical switch. A synapse in a brain is an electrical switch embedded in a chemical reactor of great sensitivity.

The presence of an active *D4DR* gene in a neuron immediately identifies that neuron as a member of one of the brain's dopamine-mediated pathways. Dopamine pathways do many things, including controlling the flow of blood through the brain. A shortage of dopamine in the brain causes an indecisive and frozen personality, unable to initiate even the body's own movement. In the extreme form, this is known as Parkinson's disease. Mice with the genes for making dopamine knocked out will starve to death from sheer immobility. If a chemical that closely resembles dopamine (a dopamine agonist, in the jargon) is injected into their brains, they recover their natural arousal. An excess of dopamine in the brain, by contrast, makes a mouse highly exploratory and adventurous. In human beings, excessive dopamine may be the immediate cause of schizophrenia; and some hallucinogenic drugs work by stimulating the dopamine system. A mouse addicted to cocaine so badly that it prefers the drug to food is experiencing the release of dopamine in a part of the brain known as the nucleus acumbens. A rat in which this 'pleasure centre' is stimulated whenever it presses a lever will learn to return to press the lever again and again. But if a dopamine-blocking chemical is added to the rat's brain, the rat quickly loses interest in the lever.

In other words, to simplify grossly, dopamine is perhaps the brain's motivation chemical. Too little and the person lacks initiative and motivation. Too much and the person is easily bored and frequently seeks new adventures. Here perhaps lies the root of a difference in personality. As Dean Hamer put it, when he set out to seek the gene for thrill-seeking personalities in the mid-1990s, he was looking for the difference between Lawrence of Arabia and Queen Victoria. Since it takes many different genes to make, control, emit and receive dopamine, let alone to build the brain in the first place, nobody, least of all Hamer, expected to find a single gene controlling exclusively this aspect of personality. Nor did he expect to find that all variation in adventure-seeking is genetic, merely that there would be genetic influences at work among others.

The first genetic difference turned up in Richard Ebstein's laboratory in Jerusalem in the *D4DR* gene on chromosome 11. *D4DR* has a variable repeat sequence in the middle, a minisatellite phrase forty-eight letters in length repeated between two and eleven times. Most of us have four or seven copies of the sequence, but some people have two, three, five, six, eight, nine, ten or eleven. The larger the number of repeats, the more ineffective is the dopamine receptor at capturing dopamine. A 'long' *D4DR* gene implies a low responsiveness to dopamine in certain parts of the brain, whereas a 'short' *D4DR* gene implies a high responsiveness.

Hamer and his colleagues wanted to know if people with the long gene had different personalities from people with the short gene. This is in effect the opposite procedure from that followed by Robert Plomin on chromosome 6, where he sought to correlate an unknown gene with a known behavioural difference (in IQ). Hamer went from the gene to the trait rather than vice versa. He measured the novelty-seeking character of 124 people on a series of set personality tests and then examined their genes.

Bingo. Of the subjects Hamer tested – admittedly not a huge sample

– people with either one or two long copies of the gene (remember there are two copies of each chromosome in each cell of the adult body, one from each parent) were distinctly more novelty-seeking than people with two short copies of the gene. 'Long' genes were defined as those with six or more repeats of the minisatellite sequence. At first Hamer was worried that he might be looking at what he calls a 'chopstick' gene. The gene for blue eyes is common in people who are bad at using chopsticks, but nobody would dream of suggesting that chopstick skill is genetically determined by the gene for eye colour. It just happens that both blue eyes and chopstick incompetence correlate with non-oriental origin for a blindingly obvious non-genetic reason called culture. Richard Lewontin uses another analogy for this fallacy: the fact that people who are good at knitting tend not to have Y chromosomes (i.e., they tend to be women) does not imply that knitting is caused by a lack of Y chromosomes.

So, to rule out a spurious correlation of this kind, Hamer repeated the study in the United States with members of one family. Again he found a clear correlation: the novelty-seekers were much more likely to have one or more copy of the long gene. This time the chopstick argument looks increasingly untenable, because any differences within a family are less likely to be cultural ones. The genetic difference may indeed contribute to the personality difference.

The argument goes like this. People with 'long' $D4DR$ genes have low responsiveness to dopamine, so they need to take a more adventurous approach to life to get the same dopamine 'buzz' that short-gened people get from simple things. In search of these buzzes they develop novelty-seeking personalities. Hamer went on to demonstrate a striking example of what it means to be a novelty seeker. Among heterosexual men, those with the long $D4DR$ genes are six times more likely to have slept with another man than those with the short genes. Among homosexual men, those with the long genes are five times more likely to have slept with a woman than those with the short genes. In both groups, the long-gened people had more sexual partners than the short-gened people.[1]

We all know people who will try anything, and conversely people who are set in their ways and reluctant to experiment with something new. Perhaps the first lot have long $D4DR$ genes and the second lot have short ones. It is not quite that simple. Hamer claims to explain no more than four per cent of novelty seeking by reference to this one gene. He estimates that novelty seeking is about forty per cent heritable, and that there are about ten equally important genes whose variation matches the variation in personality. That is just one element in personality, but there are many others, perhaps a dozen. Making the wild assumption that they all involve similar numbers of genes leads to the conclusion that there may be 500 genes that vary in tune with human personalities. These are just the ones that vary. There may be many others that do not normally vary, but if they did would affect personality.

This is the reality of genes for behaviour. Do you see now how unthreatening it is to talk of genetic influences over behaviour? How ridiculous to get carried away by one 'personality gene' among 500? How absurd to think that, even in a future brave new world, somebody might abort a foetus because one of its personality genes is not up to scratch – and take the risk that on the next conception she would produce a foetus in which two or three other genes were of a kind she does not desire? Do you see now how futile it would be to practise eugenic selection for certain genetic personalities, even if somebody had the power to do so? You would have to check each of 500 genes one by one, deciding in each case to reject those with the 'wrong' gene. At the end you would be left with nobody, not even if you started with a million candidates. We are all of us mutants. The best defence against designer babies is to find more genes and swamp people in too much knowledge.

Meanwhile, the discovery that personality has a strong genetic component can be used in some very non-genetic therapy. When naturally shy baby monkeys are fostered to confident monkey mothers, they quickly outgrow their shyness. It is almost certainly the same with people – the right kind of parenting can alter an innate personality. Curiously, understanding that it is innate seems to help to cure it. One trio of therapists, reading about the new results emerging from genetics, switched from trying to treat their clients' shyness to trying to make them content with whatever their innate predispositions were. They found that it worked. The clients felt relieved to be told that their personality was a real, innate part of them and not just a bad habit they had got into. 'Paradoxically, depathologising people's fundamental inclinations and giving group members permission to be the way they are seemed to constitute the best insurance that their self-esteem and interpersonal effectiveness would improve.' In other words, telling them they were naturally shy helped them overcome that shyness. Marriage counsellors, too, report good results from encouraging their clients to accept that they cannot change their partners' irritating habits – because they are probably innate – but must find ways to live with them. The parents of a homosexual are generally more accepting when they believe that homosexuality is an immutable part of nature rather than a result of some aspect of their parenting. Far from being a sentence, the realisation of innate personality is often a release.[2]

Suppose you wished to breed a strain of fox or rat that was more tame and less instinctively timid than the average. One way to do so would be to pick the darkest pups in the litter as the stock for breeding the next generation. In a few years you would have tamer, and darker, animals. This curious fact has been known to animal breeders for many years. But in the 1980s it took on a new significance. It parallels another link between neurochemistry and personality in people. Jerome Kagan, a Harvard psychologist, leading a team of researchers studying shyness or confidence in children, found that he could identify unusually 'inhibited' types as early as four months of age – and fourteen years later could predict how shy or confident those same human beings would be as adults. Upbringing mattered a good deal. But intrinsic personality played just as big a role.

Big deal. Nobody, except perhaps the most die-hard social determinist, would find an innate component of shyness surprising. But it turned out that the same personality traits correlated with some unexpected other features. Shy adolescents were more likely to be blue-eyed (all the subjects were of European descent), susceptible to allergies, tall and thin, narrow-faced, to have more heat-generating activity under the right forehead and a faster heartbeat, than the less shy individuals. All of these features are under the control of a particular set of cells in the embryo called the neural crest, from which a particular part of the brain, the amygdala, derives. They also all use the same neurotransmitter, called norepinephrine, a substance very like dopamine. All these features are also characteristic of northern Europeans, Nordic types for the most part. Kagan's argument goes that the Ice Age selected those better able to withstand cold in these parts: people with high metabolic rates. But a high metabolic rate is produced by an active norepinephrine system in the amygdala, and brings with it lots of different baggage – a phlegmatic and shy personality being one aspect and a pale appearance being another. Just as in foxes and rats, shy and suspicious types are paler than bold types.[3]

If Kagan is right, tall, thin adults with blue eyes are slightly more likely to become anxious when challenged than other people. An up-to-date recruitment consultant might find this handy in his headhunting. After all, employers already seek to discriminate between personalities. Most job advertisements require candidates with 'good interpersonal skills' – something that is probably partly innate. Yet it would plainly be a repellent world in which we were picked for jobs on the basis of our eye colour. Why? Physical discrimination is so much less acceptable than psychological. Yet psychological discrimination is just chemical discrimination. It is just as material as any other discrimination.

Dopamine and norepinephrine are so-called monoamines. Their close cousin, another monoamine found in the brain, is serotonin, which is also a chemical manifestation of personality. But serotonin is more complicated than dopamine and norepinephrine. It is remarkably hard to pin down its characteristics. If you have unusually high levels of serotonin in your brain you will probably be a compulsive person, given to tidiness and caution, even to the point of being neurotic about it. People with the pathological condition known as obsessive–compulsive disorder can usually alleviate their symptoms by lowering their serotonin levels. At the other end of the spectrum, people with unusually low serotonin levels in their brains tend to be impulsive. Those who commit impulsive violent crimes, or suicide, are often those with less serotonin.

Prozac works by affecting the serotonin system, though there is still controversy about exactly how it does so. The conventional theory put forward by scientists at Eli Lilly, where the drug was invented, is that Prozac inhibits the reabsorption of serotonin into neurons, and thus increases the amount of serotonin in the brain. Increased serotonin alleviates anxiety and depression and can turn even fairly ordinary people into optimists. But it remains possible that Prozac has exactly the opposite effect: that it interferes with the responses of neurons to serotonin. There is a gene on chromosome 17, called the serotonin-transporter gene, which varies, not in itself, but in the length of an 'activation sequence' just upstream of the gene – a sort of dimmer switch at the beginning of the gene, in other words, designed to slow down the expression of the gene itself. As with so many mutations, the variation in length is caused by a variable number of repetitions of the same sequence, a twenty-two-letter phrase that is repeated either fourteen or sixteen times. About one in three of us have two copies of the long sequence, which is marginally worse at switching off its gene. As a result such people have more serotonin transporter, which means that more serotonin gets carried about. These people are much less likely to be neurotic, and slightly more likely to be agreeable than the average person, whatever their sex, race, education or income.

From this, Dean Hamer concludes that serotonin is the chemical that abets, rather than alleviates, anxiety and depression. He calls it the brain's punishment chemical. Yet all sorts of evidence points in the other direction: that you feel better with more serotonin, not less. There is, for instance, a curious link between winter, a desire for snacks, and sleepiness. In some people – probably once more a genetic minority, though no gene version has yet been found that correlates with susceptibility to this condition – the dark evenings of winter lead to a craving for carbohydrate snacks in the late afternoon. Such people often need more sleep in winter, though they find their sleep less refreshing. The explanation seems to be that the brain starts making melatonin, the hormone that induces sleep, in response to the early evening darkness of winter days. Melatonin is made from serotonin, so serotonin levels drop as it gets used up in melatonin manufacture. The quickest way to raise serotonin levels again is to send more tryptophan into the brain, because serotonin is made from tryptophan. The quickest way to send more tryptophan into the brain is to secrete insulin from the pancreas, because insulin causes the body to absorb other chemicals similar to tryptophan, thus removing competitors for the channels that take tryptophan into the brain. And the quickest way to secrete insulin is to eat a carbohydrate snack.[4]

Are you still with me? You eat cookies on winter evenings to cheer yourself up by raising your brain serotonin. The take-home message is that you can alter your serotonin levels by altering your eating habits. Indeed, even drugs and diets designed to lower blood cholesterol can influence serotonin. It is a curious fact that nearly all studies of cholesterol-lowering drugs and diets in ordinary people show an increase in violent death compared with control samples that usually matches the decrease in deaths from heart disease. In all studies put together, cholesterol treatment cut heart attacks by fourteen per cent, but raised violent deaths by an even more significant seventy-eight per cent. Because violent deaths are rarer than heart attacks, the numerical effect roughly cancels out, but violent deaths can sometimes involve innocent bystanders. So treating high cholesterol levels has its dangers. It has been known for twenty years that impulsive, antisocial and depressed people – including prisoners, violent offenders and failed suicides – have generally lower cholesterol levels than the population at large. No wonder Julius Caesar distrusted Cassius's lean and hungry look.

These disturbing facts are usually played down by the medical profession as statistical artefacts, but they are too repeatable for that. In the so-called MrFit trial, in which 351,000 people from seven countries were followed for seven years, people with very low cholesterol and people with very high cholesterol proved twice as likely to die at a given age as people with medium cholesterol. The extra deaths among low-cholesterol people are mainly due to accident, suicide or murder. The twenty-five per cent of men with the lowest cholesterol count are four times as likely to commit suicide as the twenty-five per cent of men with the highest count – though no such pattern holds with women. This does not mean we should all go back to eating fried eggs. Having low cholesterol, or lowering your cholesterol too far, is highly dangerous for a small minority, just as having high cholesterol and eating high-cholesterol diets is dangerous for a small minority. Low-cholesterol dieting advice should be confined to those who are genetically endowed with too much cholesterol, and not given to everybody.

The link between low cholesterol and violence almost certainly involves serotonin. Monkeys fed on low-cholesterol diets become more aggressive and bad-tempered (even if they are not losing weight), and the cause seems to be a drop in serotonin levels. In Jay Kaplan's laboratory at Bowman Gray Medical School in North Carolina, eight monkeys fed on a low-cholesterol (but high-fat) diet soon had brain serotonin levels that were roughly half as high as those in the brains of nine monkeys fed on a high-cholesterol diet. They were also forty per cent more likely to take aggressive or antisocial action against a fellow monkey. This was true of both sexes. Indeed, low serotonin is an accurate predictor of aggressiveness in monkeys, just as it is an accurate predictor of impulsive murder, suicide, fighting or arson in human beings. Does this mean that if every man was forced by law to have his serotonin level displayed on his forehead at all times, we could tell who should be avoided, incarcerated or protected from themselves?[5]

Fortunately, such a policy is as likely to fail as it is offensive to civil liberties. Serotonin levels are not innate and inflexible. They are themselves the product of social status. The higher your self-esteem and social rank relative to those around you, the higher your serotonin level is. Experiments with monkeys reveals that it is the social behaviour that comes first. Serotonin is richly present in dominant monkeys and much more dilute in the brains of subordinates. Cause or effect? Almost everbody assumed the chemical was at least partly the cause: it just stands to reason that the dominant behaviour results from the chemical, not vice versa. It turns out to be the reverse: serotonin levels respond to the monkey's perception of its own position in the hierarchy, not vice versa.[6]

Contrary to what most people think, high rank means low aggressiveness, even in vervet monkeys. The high-ranking individuals are not especially large, fierce or violent. They are good at things like reconciliation and recruiting allies. They are notable for their calm demeanour. They are less impulsive, less likely to misinterpret play-fighting as aggression. Monkeys are not people, of course, but as Michael McGuire of the University of California, Los Angeles, has discovered, any group of people, even children, can immediately spot which of the monkeys in his captive group is the dominant one. Its demeanour and behaviour – what Shelley called the 'sneer of cold command' – are instantly familiar in an anthropomorphic way. There is little doubt that the monkey's mood is set by its high serotonin levels. If you artificially reverse the pecking order so that the monkey is now a subordinate, not only does its serotonin drop, but its behaviour changes, too. Moreover, much the same seems to happen in human beings. In university fraternities, the leading figures are blessed with rich serotonin concentrations which fall if they are deposed. Telling people they have low or high serotonin levels could become a self-fulfilling prophecy.

This is an intriguing reversal of the cartoon picture of biology most people have. The whole serotonin system is about biological determinism. Your chances of becoming a criminal are affected by your brain chemistry. But that does not mean, as it is usually assumed to mean, that your behaviour is socially immutable. Quite the reverse: your brain chemistry is determined by the social signals to which you are exposed. Biology determines behaviour yet is determined by society. I described the same phenomenon in the cortisol system of the body; here it is again with the serotonin system of the brain. Mood, mind, personality and behaviour are indeed socially determined, but that does not mean they are not also biologically determined. Social influences upon behaviour work through the switching on and off of genes.

None the less, it is clear that there are all sorts of innate personality types, and that people vary in the way they respond to social stimuli mediated through neurotransmitters. There are genes that vary the rate of serotonin manufacture, genes that vary the responsiveness of serotonin receptors, genes that make some brain areas respond to serotonin more than others, genes that make some people depressed in winter because of too responsive a melatonin system using up serotonin. And so on and on and on. There is a Dutch family in which the men have been criminals for three generations, and the cause is undoubtedly a gene. The criminal men have an unusual version of a gene on the X chromosome called the mono-amine oxidase A gene. Monoamine oxidase is responsible for breaking down serotonin among other chemicals. It is highly probable that their unusual serotonin neurochemistry makes these Dutch men more likely to fall into lives of crime. But this does not make this gene a 'crime gene', except in a very pedestrian sense. For a start, the mutation in question is now considered an 'orphan' mutation, so rare that very few criminals have this version of the gene. The monoamine oxidase gene can explain very little about general criminal behaviour.

But it underscores yet again the fact that what we call personality is to a considerable degree a question of brain chemistry. There are a score of different ways in which this one chemical, serotonin, can be related to innate differences in personality. These are overlaid on the score of different ways that the mind's serotonin system responds to outside influences such as social signals. Some people are more sensitive to some outside signals than others. This is the reality of genes and environments: a maze of complicated inter-actions between them, not a one-directional determinism. Social behaviour is not some external series of events that takes our minds and bodies by surprise. It is an intimate part of our make-up, and our genes are programmed not only to produce social behaviour, but to respond to it as well.

CHROMOSOME 1 2

Self-Assembly

The egg's ordain'd by nature to that end
And is a chicken in potentia
Ben Jonson, The Alchemist

There are human analogies for almost everything in nature. Bats use sonar; the heart is a pump; the eye is a camera; natural selection is trial and error; genes are recipes; the brain is made from wires (known as axons) and switches (synapses); the hormonal system uses feedback control like an oil refinery; the immune system is a counter-espionage agency; bodily growth is like economic growth. And so, infinitely, on. Although some of these analogies can mislead, we are at least familiar with the kinds of techniques and technologies that Mother Nature employs to solve her various problems and achieve her ingenious designs. We have reinvented most of them ourselves in technological life.

But now we must leave such comfortable terrain behind and step into the unknown. One of the most remarkable, beautiful and bizarre things that Mother Nature achieves without apparent difficulty is something for which we have no human analogy at all: the development of a human body from an undifferentiated blob called a fertilised egg. Imagine trying to design a piece of hardware (or software, for that matter) that could do something analogous to this feat. The Pentagon probably tried it, for all I know: 'Good Morning, Mandrake. Your job is to make a bomb that grows itself from a large blob of raw steel and a heap of explosive. You have an unlimited budget and one thousand of the best brains at your disposal in the New Mexico desert. I want to see a prototype by August. Rabbits can do it ten times a month. So it cannot be that hard. Any questions?'

Without the handrail of analogy, it is difficult even to understand Mother Nature's feat. Something, somewhere must be imposing a pattern of increasing detail upon the egg as it grows and develops. There must be a plan. But unless we are to invoke divine intervention, that imposer of detail must be within the egg itself. And how can the egg make a pattern without starting with one? Little wonder that, in past centuries, there was a natural preference for theories of preformation, so that some people thought they saw within the human sperm a miniature homunculus of a man. Preformation, as even Aristotle spotted, merely postpones the problem, for how did the homunculus get its shape? Later theories were not much better, though our old friend William Bateson came surprisingly close to the right answer when he conjectured that all organisms are made from an orderly series of parts or segments, and coined the term homeosis for it. And there was a vogue in the 1970s for explaining embryology by reference to increasingly sophisticated mathematical geometries, standing waves and other such arcana. Alas for mathematicians, nature's answer turns out, as ever, to be both simpler and much more easily understood, though the details are ferociously intricate. It all revolves around genes, which do indeed contain the plan in digital form. One large cluster of these developmental genes lies close to the middle of chromosome 12. The discovery of these genes and the elucidation of how they work is probably the greatest intellectual prize that modern genetics has won since the code itself was cracked. It was a discovery with two stunning and lucky surprises at its heart.[1]

As the fertilised egg grows into an embryo, at first it is an undifferentiated blob. Then gradually it develops two asymmetries – a head–tail axis and a front–back axis. In fruit flies and toads, these axes are established by the mother, whose cells instruct one end of the embryo to become the head and one part to become the back. But in mice and people the asymmetries develop later and nobody knows quite how. The moment of implantation into the womb seems to be critical.

In fruit flies and toads, these asymmetries are well understood: they consist of gradients in the chemical products of different maternal genes. In mammals, too, the asymmetries are almost certainly chemical. Each cell can, as it were, taste the soup inside itself, feed the information into its hand-held GPS microcomputer and get out a reading: 'you are in the rear half of the body, close to the underside.' Very nice to know where you are.

But knowing where you are is just the beginning. Knowing what you have to do once you are there is a wholly different problem. Genes that control this process are known as 'homeotic' genes. For instance, our cell, on discovering where it is located, looks this location up in its guidebook and finds the instruction: 'grow a wing', or 'start to become a kidney cell' or something like that. It is not of course literally like this. There are no computers and no guidebooks, just a series of automatic steps in which gene switches on gene which switches on gene. But a guidebook is a handy analogy, none the less, because the great beauty of embryo development, the bit that human beings find so hard to grasp, is that it is a totally decentralised process. Since every cell in the body carries a complete copy of the genome, no cell need wait for instructions from authority; every cell can act on its own information and the signals it receives from its neighbours. We do not organise societies that way: we are obsessed with dragging as many decisions as possible to the centre to be taken by governments. Perhaps we should try.[2]

Fruit flies have been a favourite object of geneticists' studies since the early years of the century, for they breed quickly and easily in the laboratory. It is the humble fruit fly we must thank for the elucidation of many of the basic principles of genetics: the idea that genes are linked on chromosomes, or Muller's discovery that genes can be mutated by X-rays. Among the mutant flies thus created, scientists began to find ones that had grown in unusual ways. They had legs where they should have antennae, or wings where they should have small stabilisers called halteres. A certain segment of the body, in other words, had done something appropriate to a different segment of the body. Something had gone wrong with the homeotic genes.

In the late 1970s, two scientists working in Germany named Jani Nüsslein-Volhard and Eric Wieschaus set out to find and describe as many such mutant flies as possible. They dosed the flies with chemicals that cause mutations, bred them by the thousand and slowly sorted out all the ones with limbs or wings or other body parts that grew in the wrong places. Gradually they began to see a consistent pattern. There were 'gap' genes that had big effects, defining whole areas of the body, 'pair-rule' genes that subdivided these areas and defined finer details, and 'segment-polarity' genes that subdivided those details by affecting just the front or rear of a small section. The developmental genes seemed, in other words, to act hierarchically, parcelling up the embryo into smaller and smaller sections to create ever more detail.[3]

This came as a great surprise. Until then, it had been assumed that the parts of the body defined themselves according to their neighbouring parts, not according to some grand genetic plan. But when the fruit-fly genes that had been mutated were pinned down and their sequences read, a further surprise was in store. The result was the first of two almost incredible discoveries, which between them amount to one of the most wonderful additions to knowledge of the twentieth century. The scientists found a cluster of eight homeotic genes lying together on the same chromosome, genes which became known as Hox genes. Nothing strange about that. What was truly strange was that each of the eight genes affected a different part of the fly and they were lined up *in the same order as the part of the fly they affected.* The first gene affected the mouth, the second the face, the third the top of the head, the fourth the neck, the fifth the thorax, the sixth the front half of the abdomen, the seventh the rear half of the abdomen, and the eighth various other parts of the abdomen. It was not just that the first genes defined the head end of the fly and the last genes made the rear end of the fly. They were all laid out in order along the chromosome – without exception.

To appreciate how odd this was, you must know how random the order of genes usually is. In this book, I have told the story of the genome in a sort of logical order, picking genes to suit my purpose chapter by chapter. But I have deceived you a little in doing this: there is very little rhyme or reason for where a gene lies. Sometimes it needs to be close to certain other genes. But it is surely rather literal of Mother Nature to lay these homeotic genes out in the order of their use.

A second surprise was in store. In 1983 a group of scientists working in Walter Gehring's laboratory in Basel discovered something common to all these homeotic genes. They all had the same 'paragraph' of text, 180 'letters' long, within the gene – known as the homeobox. At first, this seemed irrelevant. After all, if it was the same in every gene, it could not tell the fly to grow a leg rather than an antenna. All electrical appliances have plugs, but you cannot tell a toaster from a lamp by looking at the plug. The analogy between a homeobox and a plug is quite close: the homeobox is the bit by which the protein made by the gene attaches to a strand of DNA to switch on or off another gene. All homeotic genes are genes for switching other genes on or off.

But the homeobox none the less enabled geneticists to go looking for other homeotic genes, like a tinker rooting through a pile of junk in search of anything with a plug attached. Gehring's colleague Eddie de Robertis, acting on no more than a hunch, went fishing among the genes of frogs for a 'paragraph' that looked like the homeobox. He found it. When he looked in mice, there it was again: almost exactly the same 180-letter string – the homeobox. Not only that, the mouse also turned out to have clusters of Hox genes (four of them, rather than one) and, in the same way as the fruit fly, the genes in the clusters were laid out end-to-end with the head genes first and the tail genes last.

The discovery of mouse–fly homology was bizarre enough, implying as it does that the mechanism of embryonic development requires the genes to be in the same order as the body parts. What was doubly strange was that the mouse genes were recognisably the same genes as the fruit-fly genes. Thus the first gene in the fruit-fly cluster, called *lab*, is very similar to the first gene in each of three mouse clusters, called *a1*, *b1* and *d1*, and the same applies to each of the other genes.[4]

There are differences, to be sure. Mice have thirty-nine Hox genes altogether, in four clusters, and they have up to five extra Hox genes at the rear end of each cluster that flies do not have. Various genes are missing in each cluster. But the similarity is still mind-blowing. It was so mind-blowing when it first came to light that few embryologists believed it. There was widespread scepticism, and belief that some silly coincidence had been exaggerated. One scientist remembers that on first hearing this news he dismissed it as another of Walter Gehring's wild ideas; it soon dawned on him that Gehring was being serious. John Maddox, editor of the journal *Nature*, called it 'the most important discovery this year (so far)'. At the level of embryology we are glorified flies. Human beings have exactly the same Hox clusters as mice, and one of them, Cluster C, is right here on chromosome 12.

There were two immediate implications of this breakthrough, one evolutionary and one practical. The evolutionary implication is that we are descended from a common ancestor with flies which used the same way of defining the pattern of the embryo more than 530 million years ago, and that the mechanism was so good that all this dead creature's descendants have hung on to it. Indeed, even more different creatures, such as sea urchins, are now known to use the same gene clusters. Though a fly or a sea urchin may look very different from a person, when compared with, say, a Martian, their embryos are very similar. The incredible conservatism of embryological genetics took everybody by surprise. The practical application was that suddenly all those decades of hard work on the genes of fruit flies were of huge relevance to human beings. To this day, science knows far more about the genes of fruit flies than it knows about the genes of people. That knowledge was now doubly relevant. It was like being able to shine a bright light on the human genome.

This lesson emerges not just from Hox genes but from all developmental genes. It was once thought, with a trace of hubris, that the head was a vertebrate speciality – that we vertebrates in our superior genius invented a whole set of new genes for building a specially 'encephalised' front end, complete with brain. Now we know that two pairs of genes involved in making a brain in a mouse, *Otx* (1 and 2) and *Emx* (1 and 2), are pretty near exact equivalents of two genes that are expressed in the development of the head end of the fruit fly. A gene that is central to making eyes in the fruit fly – called oxymoronically *eyeless* – is recognisably the same as a gene that is central to making eyes in the mouse: where it is known as *pax*-6. What is true of mice is just as true of people. Flies and people are just variations on a theme of how to build a body that was laid down in some worm-like creature in the Cambrian period. They still retain the same genes doing the same job. Of course, there are differences; if there were not, we would look like flies. But the differences are surprisingly subtle.

The exceptions are almost more convincing than the rule. For instance, in flies there are two genes that are crucial to laying down the difference between the back (dorsal) of the body and the front (ventral). One, called *decapentaplegic*, is dorsalising – i.e., when expressed it makes cells become part of the back. The other, called *short gastrulation*, is ventralising – it makes cells become part of the belly. In toads, mice and almost certainly in you and me, there are two very similar genes. The 'text' of one, *BMP4*, reads very like the 'text' of *decapentaplegic*; the 'text' of the other, *chordin*, reads very like the text of *short gastrulation*. But, astonishingly, each of these has the opposite effect in mice that its equivalent has in flies: *BMP4* is ventralising, and *chordin* is dorsalising. This means that arthropods and vertebrates are upside-down versions of each other. Some time in the ancient past they had a common ancestor. And one of the descendants of the common ancestor took to walking on its stomach while the other took to walking on its back. We may never know which one was 'the right way up', but we do know that there was a right way up, because we know the dorsalising and ventralising genes predate the split between the two lineages. Pause, for a second, to pay homage to a great Frenchman, Étienne Geoffroy St Hilaire, who first guessed this fact in 1822, from observing the way embryos develop in different animals and from the fact that the central nervous system of an insect lies along its belly while that of a human being lies along its back. His bold conjecture was subjected to much ridicule in the intervening 175 years, and conventional wisdom accreted round a different hypothesis, that the nervous systems of the two kinds of animals were independently evolved. But he was absolutely right.[5]

Indeed, so close are the similarities between genes that geneticists can now do, almost routinely, an experiment so incredible that it boggles the mind. They can knock out a gene in a fly by deliberately mutating it, replace it by genetic engineering with the equivalent gene from a human being and grow a normal fly. The technique is known as genetic rescue. Human Hox genes can rescue their fly equivalents, as can *Otx* and *Emx* genes. Indeed, they work so well that it is often impossible to tell which flies have been rescued with human genes and which with fly genes.[6]

This is the culminating triumph of the digital hypothesis with which this book began. Genes are just chunks of software that can run on any system: they use the same code and do the same jobs. Even after 530 million years of separation, our computer can recognise a fly's software and vice versa. Indeed, the computer analogy is quite a good one. The time of the Cambrian explosion, between 540 and 520 million years ago, was a time of free experimentation in body design, a bit like the mid-1980s in computer software. It was probably the moment when the first homeotic genes were invented by one lucky species of animal from which we are all descended. This creature was almost certainly a mud-burrowing thing known – with delicate contradiction – as the Roundish Flat Worm, or RFW. It was probably just one of many rival body plans, but its descendants inherited the earth or large chunks thereof. Was it the best design, or just the most brilliantly marketed? Who was the Apple of the Cambrian explosion and who the Microsoft?

Let us take a closer look at one of the Hox genes on human chromosome 12. Hox $C4$ is the genetic equivalent of a gene called *dfd* in flies, which is expressed in what will become the mouthparts of the adult fly. It is also very close in sequence to its counterparts on other chromosomes, $A4$, $B4$ and $D4$ – and the mouse versions of the same genes: $a4$, $b4$, $c4$ and $d4$. In the embryo of a mouse, these genes are expressed in the part that will become the neck: the cervical vertebrae and the spinal cord within them. If you 'knock out' one of these genes by mutation, you find that one or two of the vertebrae of the mouse's neck are affected. But the effect of the knock-out is very specific. It makes the affected vertebrae grow as if they were further forward in the mouse's neck than they are. The Hox 4 genes are needed to make each neck vertebra different from the first neck vertebra. If you knock out two of the Hox 4 genes, more vertebrae are affected, and if you knock out three of the four genes, even more cervical vertebrae are affected. Therefore, the four genes seem to have a sort of cumulative effect. Moving from head to rear, the genes are switched on one after another and each new gene turns that part of the embryo into a more posterior body part. By having four versions of each Hox gene, we and mice have rather more subtle control over the development of our bodies than flies do with just one Hox cluster.

It also becomes clear why we have up to thirteen Hox genes in each cluster rather than eight, as flies do. Vertebrates have post-anal tails, that is spines which go on well past their anuses. Insects do not. The extra Hox genes that mice and people have, which flies do not, are needed for programming the development of the lower back and tail. Since our ancestors, when they became apes, shrank their tails to nothing, these genes are presumably somewhat silent in us compared with their equivalents in mice.

We are now in a position to face a vital question. Why are the Hox genes laid end-to-end, with the first genes expressed at the head of the animal, in every species so far investigated? There is as yet no definitive answer, but there is an intriguing hint. The foremost gene to be expressed is not only expressed in the foremost part of the body; it is also the first to be expressed. All animals develop from the bow to the stern. So the co-linear expression of the Hox genes follows a temporal sequence, and it is probable that the switching on of each Hox gene somehow switches on the next one in line or allows it to be opened up and read. Moreover, the same is probably true of the animal's evolutionary history. Our ancestors seem to have grown more complicated bodies by lengthening and developing the rear end, not the head end. So the Hox genes replay an ancient evolutionary sequence. In Ernst Haeckel's famous phrase, 'ontogeny recapitulates phylogeny'. The embryo's development occurs in the same sequence as its ancestors' evolution.[7]

Neat as these tales are, they tell only a fraction of the story. We have given the embryo a pattern – a top–down asymmetry and a bow–stern asymmetry. We have given it a set of genes that get turned on according to a clever sequence of timing and thus are each expressed in a different part of the body. Each Hox compartment has switched on its special Hox gene, which in turn has switched on other genes. The compartment must now differentiate in the appropriate way. It must, for example, grow a limb. The clever part of what happens next is that the same signals are now used to mean different things in different parts of the body. Each compartment knows its location and identity and reacts to the signals accordingly. Our old friend *decapentaplegic* is one of the triggers for the development of a leg in one compartment of a fly and a wing in another. It in turn is triggered by another gene called *hedgehog*, whose job is to interfere with the proteins that keep *decapentaplegic* silenced and thus to awaken it. *Hedgehog* is a so-called segment-polarity gene, which means it is expressed in every segment, but only in the rear half thereof. So if you move a *hedgehog*-expressing piece of tissue into the anterior half of the wing segment, you get a fly with a sort of mirror-image wing with two front halves fused back to back in the middle and two back halves on the outsides.

It will not surprise you to learn that *hedgehog* has its equivalents in people and in birds. Three very similar genes, called *sonic hedgehog*, *Indian hedgehog* and *desert hedgehog*, do much the same thing in chicks and people. (I told you geneticists had strange minds: there is now a gene called *tiggywinkle* and two new gene families called *warthog* and *groundhog*. It all started because fruit flies with faulty *hedgehog* genes had a prickly appearance.) Just as in the fly, the job of *sonic hedgehog* and its scheming partners is to tell the compartment where the rear half of the limb should be. It is switched on when a blunt limb bud has already formed, telling the limb bud which way is rear. If at the right moment you take a microscopic bead, soak it in *sonic hedgehog* protein and insert it carefully into the thumb side of the wing bud of a chick embryo for twenty-four hours, the result will be two mirror-image wings fused front half to front half and with two back halves on the outsides – almost precisely the same result as in fruit flies.

The *hedgehog* genes, in other words, define the front and rear of the wing, and it is Hox genes that then divide it up into digits. The transformation of a simple limb bud into a five-fingered hand happens in every one of us, but it also happened, on a different timescale, when the first tetrapods developed hands from fish fins some time after 400 million years ago. In one of the most satisfying pieces of recent science, palaeontologists studying that ancient transformation have come together with embryologists studying Hox genes and discovered common ground.

The story starts with the discovery in Greenland in 1988 of a fossil called *Acanthostega*. Half fish and half tetrapod, and dating from 360 million years ago, it surprised everybody by having typical tetrapod limbs with eight-digit hands on the end of them. It was one of several experimental limb designs tried out by the early tetrapods as they crawled through shallow water. Gradually, from other such fossils, it became clear that the hand we all possess developed in a curious way from the fish's fin: by the development of a forward-curving arch of bones in the wrist from which digits were flung off towards the rear (little-finger) side. You can still just see this pattern in an X-ray of your own hand. All this was worked out from dry bones of fossils, so imagine the palaeontologists' surprise when they read of the embryologists' discovery that this is exactly how the Hox genes go about their work in the limb. First they set up a gradient of expression curving towards the front of the growing limb, to divide it into separate arm and wrist bones, then they suddenly set up a reverse gradient on the outside of the last bones to throw off the five digits.[8]

Hox and *hedgehog* genes are not by any means the only genes that control development. Scores of other genes doing ingenious things to signal where and how bits of the body should grow make up a system of brilliant self-organisation: 'pax genes' and 'gap genes', genes with names like *radical fringe, even-skipped, fushi tarazu, hunchback, Krüppel, giant, engrailed, knirps, windbeutel, cactus, huckebein, serpent, gurken, oskar* and *tailless*. Entering the new world of genetic embryology sometimes feels like dropping into a Tolkien novel; it requires you to learn a massive vocabulary. But – and here is the wonder of it – you do not need to learn a new way of thinking. There is no fancy physics, no chaos theory or quantum dynamics, no conceptual novelties. Like the discovery of the genetic code itself, what seemed initially to be a problem that could only be solved with new concepts turns out to be just a simple, literal and easily understood sequence of events. From the basic asymmetry of chemicals injected into the egg all else follows. Genes turn each other on, giving the embryo a head and a rear. Other genes then get turned on in sequence from bow to stern giving each compartment an identity. Other genes then polarise the compartments into front and rear halves. Other genes then interpret all this information and make ever more complicated appendages and organs. It is a rather basic, chemical–mechanical, step-by-step process that would have appealed more to Aristotle than Socrates. From simple asymmetry can grow intricate pattern. Indeed, so simple is embryonic development in principle – though not in detail – that it is tempting to wonder if human engineers should not try to copy it, and invent self-assembling machines.

Pre-History

Antiquitas saeculi juventus mundi (Ancient times were the youth of the world)   *Francis Bacon*

The surprising similarity of embryological genes in worms, flies, chicks and people sings an eloquent song of common descent. The reason we know of this similarity is because DNA is a code written in a simple alphabet – a language. We compare the vocabulary of developmental genes and find the same words. On a completely different scale, but with direct analogy, the same is true of human language: by comparing the vocabularies of human languages, we can deduce their common ancestry. Italian, French, Spanish and Romanian share word roots from Latin, for instance. These two processes – linguistic philology and genetic phylogeny – are converging upon a common theme: the history of human migrations. Historians may lament the lack of written records to document the distant, prehistoric past, but there is a written record, in the genes, and a spoken one, too, in the very vocabulary of human language. For reasons that will slowly emerge, chromosome 13 is a good place to discuss the genetics of genealogy.

In 1786 Sir William Jones, a British judge in Calcutta, announced to a meeting of the Royal Asiatic Society that his studies of the archaic Indian language Sanskrit had led him to conclude that it was a cousin of Latin and Greek. Being a learned fellow he also thought he saw similarities between these three languages and Celtic, Gothic and Persian. They had all, he suggested, 'sprung from some common source'. His reasoning was exactly the same as the reasoning which led modern geneticists to propose the existence of the Roundish Flat Worm of 530 million years ago: similarities of vocabulary. For instance, the word for three is 'tres' in Latin, 'treis' in Greek and 'tryas' in Sanskrit. Of course, the great difference between spoken languages and genetic languages is that there is much more horizontal borrowing of words in spoken language. Perhaps the word for three had somehow been inserted into Sanskrit from a western tongue. But subsequent research has confirmed that Jones was absolutely right and that there was once a single people, speaking a single language in a single place and that descendants of those people brought that language to lands as far apart as Ireland and India, where it gradually diverged into modern tongues.

We can even learn something about these people. The Indo-Europeans, as they are known, expanded at least 8,000 years ago from their homeland, which some think was in the modern Ukraine, but was more likely in a hilly part of modern Turkey (the language had words for hills and fast-flowing streams). Whichever is correct, the people were undoubtedly farmers – their language also had words for crops, cows, sheep and dogs. Since this dates them to soon after the very invention of agriculture in the so-called fertile crescent of Syria and Mesopotamia, we can easily picture that their immense success in stamping their mother tongue on two continents was due to their agricultural technology. But did they impose their genes in the same way? It is a question I shall have to attack indirectly.

Today in the Indo-European homeland of Anatolia, people speak Turkish, a non-Indo-European tongue brought later by horse-riding nomads and warriors from the steppes and deserts of central Asia. These 'Altaic' people owned a superior technology, too – the horse

– and their vocabulary confirms as much: it is full of common words for horses. A third family of languages, the Uralic, spoken in northern Russia, Finland, Estonia and, bizarrely, Hungary, bears witness to a previously successful expansion of people, before and after the Indo-Europeans, using an unknown technology – herding of domestic animals, perhaps. Today the Samoyede reindeer herders of northern Russia are perhaps typical Uralic speakers. But if you delve deeper, there is undoubtedly a family connection between these three linguistic families: Indo-European, Altaic and Uralic. They derive from a single language spoken throughout Eurasia maybe 15,000 years ago by hunter-gathering people who had, to judge by the words in common in their descendant tongues, not yet domesticated any animals, except possibly the wolf (dog). There is disagreement about where to draw the boundaries that contain the descendants of these 'Nostratic' people. The Russian linguists Vladislav Illich-Svitych and Aharon Dolgopolsky prefer to include the Afro-Asiatic family of languages spoken in Arabia and North Africa, whereas Joseph Greenberg of Stanford University omits them but includes the Kamchatkan and Chukchi languages of north-east Asia. Illich-Svitych even wrote a little poem in phonetic Nostratic, having deduced what the root words sounded like.

The evidence for this linguistic super-family lies in the simple little words that change least. Indo-European, Uralic, Mongol, Chukchi and Eskimo languages, for example, almost all use or used the 'm' sound in the word for 'me' and the 't' sound in the word for 'you' (as in the French 'tu'). A string of such examples stretches to breaking point the coincidence hypothesis. Remarkable as it seems, the languages spoken in Portugal and Korea are almost certainly descended from the same single tongue.

Quite what the Nostratic people's secret was we may never know. Perhaps they had invented hunting with dogs or stringed weapons for the first time. Perhaps it was something less tangible, like democratic decision making. But they did not altogether wipe out their predecessors. There is good evidence that Basque, several languages spoken in the Caucasus mountains and now-extinct Etruscan do not belong to the Nostratic super-family of languages, but share an affinity with Navajo and some Chinese tongues in a different super-family known as Na-Dene. We are getting into highly speculative bideas here, but Basque, which survived in the Pyrenees (mountains are backwaters of human migration, bypassed by the main flows), was once spoken in a larger area, as shown by place names, and the area coincides neatly with the painted caves of Cro-Magnon hunters. Are Basque and Navajo linguistic fossils of the first modern people to oust the Neanderthals and spread into Eurasia? Are speakers of these tongues actually descended from mesolithic people, and surrounded by neighbours of neolithic descent speaking Indo-European languages? Probably not, but it is a delicious possibility.

In the 1980s Luigi Luca Cavalli-Sforza, a distinguished Italian geneticist, watched these unfolding discoveries of linguistics and decided to ask the obvious question: do linguistic boundaries coincide with genetic ones? Genetic boundaries are inevitably more blurred, because of intermarriage (most people speak only one language, but share the genes of four grandparents). The differences between French and German genes are much less definite than the difference between the French and the German languages.

None the less, some patterns emerge. By gathering data on the common, known variations in simple genes – the 'classical polymorphisms' – and doing clever statistical tricks called principal-components analysis with the resulting data, Cavalli-Sforza uncovered five different contour maps of gene frequencies within Europe. One was a steady gradient from south-east to north-west, which may reflect the original spread of neolithic farmers into Europe from the Middle East: it echoes almost exactly the archaeological data on the spread of agriculture into Europe beginning about 9,500 years ago. This accounts for twenty-eight per cent of the genetic variation in his sample. The second contour map was a steep hill to the north-east, reflecting the genes of the Uralic speakers, and accounting for twenty-two per cent of genetic variation. The third, half as strong, was a concentration of genetic frequencies radiating out from the Ukrainian steppes, reflecting the expansion of pastoral nomads from the steppes of the Volga–Don region in about 3,000 BC. The fourth, weaker still, peaks in Greece, southern Italy and western Turkey, and probably shows the expansion of Greek peoples in the first and second millennium BC. Most intriguing of all, the fifth is a steep little peak of unusual genes coinciding almost exactly with the greater (original) Basque country in northern Spain and southern France. The suggestion that Basques are survivors of the pre-neolithic peoples of Europe begins to seem plausible.[1]

Genes, in other words, support the evidence from linguistics that expansions and migrations of people with novel technological skills have played a great part in human evolution. The gene maps are fuzzier than the linguistic maps, but this enables them to be subtler. On a smaller scale, too, they can pick out features that coincide with linguistic regions. In Cavalli-Sforza's native Italy, for instance, there are genetic regions that coincide with the ancient Etruscans, the Ligurians of the Genoa region (who spoke a non-Indo-European ancient language) and the Greeks of southern Italy. The message is plain. Languages and peoples do, to some extent, go together.

Historians speak happily of neolithic people, or herdsmen, or Magyars, or whoever, 'sweeping into' Europe. But what exactly do they mean? Do they mean expanding, or migrating? Do these newcomers displace the people already there? Do they kill them, or merely out-breed them? Do they marry their women and kill their men? Or do their technology, language and their culture merely spread by word of mouth and become adopted by the natives? All models are possible. In the case of eighteenth-century America, the native Americans were displaced almost completely by whites – both in genetic and linguistic terms. In seventeenth-century Mexico, something much more like mixing happened. In nineteenth-century India, the language of English spread, as a whole procession of Indo-European languages such as Urdu/Hindi had done before, but in this case with very little genetic admixture.

The genetic information allows us to understand which of these models applies best to pre-history. The most plausible way to account for a genetic gradient that grows steadily more dilute towards the north-west is to imagine a spread of neolithic agriculture by diffusion. That is, the neolithic farmers from the south-east must have mixed their genes with those of the 'natives', the influence of the invaders' genes growing steadily less distinct the further they spread. This points to intermarriage. Cavalli-Sforza argues that the male cultivators probably married the local hunter-gatherer women, but not vice versa, because that is exactly what happens between the pygmies and their cultivator neighbours in central Africa today. Cultivators, who can afford more polygamy than hunter-gatherers, and tend to look down on foraging people as primitive, do not allow their own women to marry the foragers, but the male cultivators do take forager wives.

Where invading men have imposed their language upon a land but married the local women, there should be a distinct set of Y-chromosome genes but a less distinct set of other genes. This is the case in Finland. The Finns are genetically no different from the other western Europeans who surround them, except in one notable respect: they have a distinct Y chromosome, which looks much more like the Y chromosome of northern Asian people. Finland is a place where the Uralic language and the Uralic Y chromosomes were imposed on a genetically and linguistically Indo-European population some time in the distant past.[2]

What has all this to do with chromosome 13? It so happens that there is a notorious gene called *BRCA2* on chromosome 13 and it, too, helps to tell a story of genealogy. *BRCA2* was the second 'breast cancer gene' to be discovered, in 1994. People with a certain, fairly rare version of *BRCA2* were found to be much more likely to develop breast cancer than is usually the case. The gene was first located by studying Icelandic families with a high incidence of breast cancer. Iceland is the perfect genetic laboratory because it was settled by such a small group of Norwegians around AD 900, and has seen so little immigration since. Virtually all of the 270,000 Icelanders trace their descent in all lines from those few thousand Vikings who reached Iceland before the little ice age. Eleven hundred years of chilly solitude and a devastating fourteenth-century plague have rendered the island so inbred that it is a happy genetic hunting ground. Indeed, an enterprising Icelandic scientist working in America returned to his native country in recent years precisely to start a business helping people to track down genes.

Two Icelandic families with a history of frequent breast cancer can be traced back to a common ancestor born in 1711. They both have the same mutation, a deletion of five 'letters' after the 999th 'letter' of the gene. A different mutation in the same gene, the deletion of the 6,174th 'letter', is common in people of Ashkenazi Jewish descent. Approximately eight per cent of Jewish breast-cancer cases under the age of forty-two are attributable to this one mutation, and twenty per cent to a mutation in *BRCA1*, a gene on chromosome 17. Again, the concentration points to past inbreeding, though not on the Icelandic scale. Jewish people retained their genetic integrity by adding few converts to the faith and losing many people who married outsiders. As a result, the Ashkenazim in particular are a favourite people for genetic studies. In the United States the Committee for the Prevention of Jewish Genetic Disease organises the testing of schoolchildren's blood. When matchmakers are later considering a marriage between two young people, they can call a hotline and quote the two anonymous numbers they were each assigned at the testing. If they are both carriers of the same mutation, for Tay–Sachs disease or cystic fibrosis, the committee advises against the marriage. The practical results of this voluntary policy – which was criticised in 1993 by the *New York Times* as eugenic – are already impressive. Cystic fibrosis has been virtually eliminated from the Jewish population in the United States.[3]

So genetic geography is of more than academic interest. Tay–Sachs disease is the result of a genetic mutation comparatively common in Ashkenazi Jews, for reasons that will be familiar from chromosome 9. Tay–Sachs carriers are somewhat protected against tuberculosis, which reflects the genetic geography of Ashkenazi Jews. Crammed into urban ghettos for much of the past few centuries, the Ashkenazim were especially exposed to the 'white death' and it is little wonder that they acquired some genes that offer protection, even at the expense of lethal complications for a few.

Although no such easy explanation yet exists for the mutation on chromosome 13 that predisposes Ashkenazis to develop breast cancer, it is quite possible that many racial and ethnic genetic peculiarities do indeed have a reason for their existence. In other words, the genetic geography of the world has a functional as well as a mapping contribution to make to the piecing together of history and pre-history.

Take two striking examples: alcohol and milk. The ability to digest large amounts of alcohol depends to some extent on the overproduction by a certain set of genes on chromosome 4 of enzymes called alcohol dehydrogenases. Most people do have the capacity to pump up production by these genes, a biochemical trick they perhaps evolved the hard way – that is, by the death and disabling of those without it. It was a good trick to learn, because fermented liquids are relatively clean and sterile. They do not carry germs. The devastation wrought by various forms of dysentery in the first millennia of settled agricultural living must have been terrible. 'Don't drink the water', we westerners tell each other when heading for the tropics. Before bottled water, the only supply of safe drinking water was in boiled or fermented form. As late as the eighteenth century in Europe, the rich drank nothing but wine, beer, coffee and tea. They risked death otherwise. (The habit dies hard.)

But foraging, nomadic people not only could not grow the crops to ferment; they did not need the sterile liquid. They lived at low densities and natural water supplies were safe enough. So it is little wonder that the natives of Australia and North America were and are especially vulnerable to alcoholism and that many cannot now 'hold their drink'.

A similar story is taught by a gene on chromosome 1, the gene for lactase. This enzyme is necessary for the digestion of lactose, a sugar abundant in milk. We are all born with this gene switched on in our digestive system, but in most mammals – and therefore in most people – it switches off during infancy. This makes sense: milk is something you drink in infancy and it is a waste of energy making the enzyme after that. But some few thousand years ago, human beings hit on the underhand trick of stealing the milk from domestic animals for themselves, and so was born the dairy tradition. This was fine for the infants, but for adults, the milk proved difficult to digest in the absence of lactase. One way round the problem is to let bacteria digest the lactose and turn the milk into cheese. Cheese, being low in lactose, is easily digestible for adults and children.

Occasionally, however, the control gene which switches off the lactase gene must suffer a mutation and the lactase production fails to cease at the end of infancy. This mutation allows its carrier to drink and digest milk all through life. Fortunately for the makers of Corn Flakes and Weetabix, most western people have acquired the mutation. More than seventy per cent of western Europeans by descent can drink milk as adults, compared with less than thirty per cent of people from parts of Africa, eastern and south-eastern Asia and Oceania. The frequency of this mutation varies from people to people and place to place in a fine and detailed pattern, so much so that it enables us to pose and answer a question about the reason people took up milk drinking in the first place.

There are three hypotheses to consider. First and most obvious, people took up milk drinking to provide a convenient and sustainable supply of food from herds of pastoral animals. Second, they took up milk drinking in places where there is too little sunlight and there is therefore a need for an extra source of vitamin D, a substance usually made with the help of sunlight. Milk is rich in vitamin D. This hypothesis was sparked by the observation that northern Europeans traditionally drink raw milk, whereas Mediterranean people eat cheese. Third, perhaps milk drinking began in dry places where water is scarce, and was principally an extra source of water for desert dwellers. Bedouin and Tuareg nomads of the Saharan and Arabian deserts are keen milk drinkers, for example.

By looking at sixty-two separate cultures, two biologists were able to decide between these theories. They found no good correlation between the ability to drink milk and high latitudes, and no good correlation with arid landscapes. This weakens the second and third hypotheses. But they did find evidence that the people with the highest frequency of milk-digestion ability were ones with a history of pastoralism. The Tutsi of central Africa, the Fulani of western Africa, the Bedouin, Tuareg and Beja of the desert, the Irish, Czech and Spanish people – this list of people has almost nothing in common except that all have a history of herding sheep, goats or cattle. They are the champion milk digesters of the human race.[4]

The evidence suggests that such people took up a pastoral way of life first, and developed milk-digesting ability later in response to it. It was not the case that they took up a pastoral way of life because they found themselves genetically equipped for it. This is a significant discovery. It provides an example of a cultural change leading to an evolutionary, biological change. The genes can be induced to change by voluntary, free-willed, conscious action. By taking up the sensible lifestyle of dairy herdsmen, human beings created their own evolutionary pressures. It almost sounds like the great Lamarckian heresy that bedevilled the study of evolution for so long: the notion that a blacksmith, having acquired beefy arms in his lifetime, then had children with beefy arms. It is not that, but it is an example of how conscious, willed action can alter the evolutionary pressures on a species – on our species in particular.

Immortality

Heaven from all creatures hides the book of fate,
All but the page prescribed, their present state.
Alexander Pope, An Essay on Man

Looking back from the present, the genome seems immortal. An unbroken chain of descent links the very first ur-gene with the genes active in your body now – an unbroken chain of perhaps fifty billion copyings over four billion years. There were no breaks or fatal mistakes along the way. But past immortality, a financial adviser might say, is no guarantee of future immortality. Becoming an ancestor is difficult – indeed, natural selection requires it to be difficult. If it were easy, the competitive edge that causes adaptive evolution would be lost. Even if the human race survives another million years, many of those alive today will contribute no genes to those alive a million years hence: their particular descendants will peter out in childlessness. And if the human race does not survive (most species last only about ten million years and most leave no descendant species behind: we've done five million years and spawned no daughter species so far), none of us alive today will contribute anything genetic to the future. Yet so long as the earth exists in something like its present state, some creature somewhere will be an ancestor of future species and the immortal chain will continue.

If the genome is immortal, why does the body die? Four billion years of continuous photocopying has not dulled the message in your genes (partly because it is digital), yet the human skin gradually loses its elasticity as we age. It takes fewer than fifty cell doublings to make a body from a fertilised egg and only a few hundred more to keep the skin in good repair. There is an old story of a king who promised to reward a mathematician for some service with anything he wanted. The mathematician asked for a chessboard with one grain of rice on the first square, two on the second, four on the third, eight on the fourth and so on. By the sixty-fourth square, he would need nearly twenty million million million grains of rice, an impossibly vast number. Thus it is with the human body. The egg divides once, then each daughter cell divides again, and so on. In just forty-seven doublings, the resulting body has more than 100 trillion cells. Because some cells cease doubling early and others continue, many tissues are created by more than fifty doublings, and because some tissues continue repairing themselves throughout life, certain cell lines may have doubled several hundred times during a long life. That means their chromosomes have been 'photocopied' several hundred times, enough to blur the message they contain. Yet fifty billion copyings since the dawn of life did not blur the genes you inherited. What is the difference?

Part of the answer lies on chromosome 14, in the shape of a gene called *TEP1*. The product of *TEP1* is a protein, which forms part of a most unusual little biochemical machine called telomerase. Lack of telomerase causes, to put it bluntly, senescence. Addition of telomerase turns certain cells immortal.

The story starts with a chance observation by James Watson, DNA's co-discoverer, in 1972. Watson noticed that the biochemical machines that copy DNA, called polymerases, cannot start at the very tip of a DNA strand. They need to start several 'words' into the text. Therefore the text gets a little shorter every time it is duplicated. Imagine a photocopier that makes perfect copies of your text but always starts with the second line of each page and ends with the penultimate line. The way to cope with such a maddening machine would be to start and end each page with a line of repeated nonsense that you do not mind losing. This is exactly what chromosomes do. Each chromosome is just a giant, supercoiled, foot-long DNA molecule, so it can all be copied except the very tip of each end. And at the end of the chromosome there occurs a repeated stretch of meaningless 'text': the 'word' TTAGGG repeated again and again about two thousand times. This stretch of terminal tedium is known as a telomere. Its presence enables the DNA-copying devices to get started without cutting short any sense-containing 'text'. Like the little plastic bit on the end of a shoelace, it stops the end of the chromosome from fraying.

But every time the chromosome is copied, a little bit of the telomere is left off. After a few hundred copyings, the chromosome is getting so short at the end that meaningful genes are in danger of being left off. In your body the telomeres are shortening at the rate of about thirty-one 'letters' a year – more in some tissues. That is why cells grow old and cease to thrive beyond a certain age. It may be why bodies, too, grow old – though there is fierce disagreement on this point. In an eighty-year-old person, telomeres are on average about five-eighths as long as they were at birth.[1]

The reason that genes do not get left off in egg cells and sperm cells, the direct ancestors of the next generation, is the presence of telomerase, whose job is to repair the frayed ends of chromosomes, re-lengthening the telomeres. Telomerase, discovered in 1984 by Carol Greider and Elizabeth Blackburn, is a curious beast. It contains RNA, which it uses as a template from which to rebuild telomeres, and its protein component bears a striking resemblance to reverse transcriptase, the enzyme that makes retroviruses and transposons multiply within the genome (see the chapter on chromosome 8). Some think it is the ancestor of all retroviruses and transposons, the original inventor of RNA-to-DNA transcription. Some think that because it uses RNA, it is a relic of the ancient RNA world.[2]

In this context, note that the 'phrase' TTAGGG, which is repeated a few thousand times in each telomere, is exactly the same in the telomeres of all mammals. Indeed, it is the same in most animals, and even in protozoans, such as the trypanosome that causes sleeping sickness, and in fungi such as *Neurospora*. In plants the phrase has an extra T at the beginning: TTTAGGG. The similarity is too close to be coincidental. Telomerase has been around since the dawn of life, it seems, and has used almost the same RNA template in all descendants. Curiously, however, the ciliate protozoans – busy microscopic creatures covered in self-propelling fur – stand out as having a somewhat different phrase repeated in their telomeres, usually TTTTGGGG or TTGGGG. The ciliates, you may remember, are the organisms that most frequently diverge from the otherwise-universal genetic code. More and more evidence points to the conclusion that the ciliates are peculiar creatures that do not fit easily into the files of life. It is my personal gut feeling that we will one day conclude that they spring from the very root of the tree of life before even the bacteria evolved, that they are, in effect, living fossils of the daughters of Luca herself, the last universal common ancestor of all living things. But I admit this is a wild surmise – and a digression.[3]

Perhaps ironically, the complete telomerase machine has been isolated only in ciliates, not in human beings. We do not yet know for sure what proteins are brought together to make up human telomerase and it may prove very different from that in ciliates. Some sceptics refer to telomerase as 'that mythical enzyme', because it is so hard to find in human cells. In ciliates, which keep their working genes in thousands of tiny chromosomes each capped with two telomeres, telomerase is much easier to find. But by searching a library of mouse DNA for sequences that resemble those used in the ciliate telomerase, a group of Canadian scientists found a mouse gene that resembled one of the ciliate genes; they then quickly found a human gene that matched the mouse gene. A team of Japanese scientists mapped the gene to chromosome 14; it produces a protein with the grand, if uncertain title of telomerase-associated protein 1, or TEP1. But it looks as if this protein, although a vital ingredient of telomerase, is not the bit that does the actual reverse transcription to repair the ends of chromosomes. A better candidate for that function has since been found but, as of this writing, its genetic location is still uncertain.[4]

Between them, these telomerase genes are as close as we may get to finding the 'genes for youth'. Telomerase seems to behave like the elixir of eternal life for cells. Geron Corporation, a company devoted to telomerase research, was founded by the scientist who first showed that telomeres shrink in dividing cells, Cal Harley. Geron hit the headlines in August 1997 for cloning part of telomerase. Its share price promptly doubled, not so much on the hope that it could give us eternal youth as on the prospect of making anti-cancer drugs: tumours require telomerase to keep them growing. But Geron went on to immortalise cells with telomerase. In one experiment, Geron scientists took two cell types grown in the laboratory, both of which lacked natural telomerase, and equipped them with a gene for telomerase. The cells continued dividing, vigorous and youthful, far beyond the point where they would normally senesce and die. At the time the result was published the cells that had had the telomerase gene introduced had exceeded their expected lifespan by more than twenty doublings, and they showed no sign of slowing down.[5]

In normal human development, the genes that make telomerase are switched off in all but a few tissues of the developing embryo. The effect of this switching off of telomerase has been likened to the setting of a stopwatch. From that moment the telomeres count the number of divisions in each cell line and at a certain point they reach their limit and call a halt. Germ cells never start the stopwatch – they never switch off the telomerase genes. Malignant tumour cells switch the genes back on. Mouse cells in which one of the telomerase genes has been artificially 'knocked out' have progressively shorter telomeres.[6]

The lack of telomerase seems to be the principal reason that cells grow old and die, but is it the principal reason bodies grow old and die? There is some good evidence in favour: cells in the walls of arteries generally have shorter telomeres than cells in the walls of veins. This reflects the harder lives of arterial walls, which are subject to more stress and strain because arterial blood is under higher pressure. They have to expand and contract with every pulse beat, so they suffer more damage and need more repair. Repair involves cell copying, which uses up the ends of telomeres. The cells start to age, which is why we die from hardened arteries, not from hardened veins.[7]

The ageing of the brain cannot be explained so easily, because brain cells do not replace themselves during life. Yet this is not fatal to the telomere theory: the brain's support cells, called glial cells, do indeed duplicate themselves; their telomeres do, therefore, probably shrink. However, there are very few experts who now believe that ageing is, chiefly, the accumulation of senescent cells, cells with abridged telomeres. Most of the things we associate with ageing – cancer, muscle weakness, tendon stiffness, hair greyness, changes in skin elasticity – have nothing to do with cells failing to duplicate themselves. In the case of cancer, the problem is that cells are copying themselves all too enthusiastically.

Moreover, there are huge differences between different species of animal in the rate at which they age. On the whole, bigger animals, such as elephants, live longer than smaller animals, which is at first sight puzzling given that it takes more cell doublings to make an elephant than a mouse – if cell doublings lead to senescent cells. And lethargic, slow-lived animals such as tortoises and sloths are long-lived for their size. This led to a neat generalisation, which is so tidy it ought to be true and probably would be if physicists ran the world: every animal has roughly the same number of heartbeats per lifetime. An elephant lives longer than a mouse, but its pulse rate is so much slower that, measured in heartbeats, they both live lives of the same length.

The trouble is, there are damning exceptions to the rule: notably bats and birds. Tiny bats can live for at least thirty years, during almost all of which they eat, breathe and pump blood at a frantic rate – and this applies even in species that do not hibernate. Birds – whose blood is several degrees hotter, whose blood sugar is at least twice as concentrated and whose oxygen consumption is far faster than in most mammals – generally live long lives. There is a famous pair of photographs of the Scottish ornithologist George Dunnet holding the same wild fulmar petrel in 1950 and 1992. The fulmar looks exactly the same in the two pictures; Professor Dunnet doesn't.

Fortunately, where the biochemists and medics have failed to explain ageing patterns, the evolutionists have come to the rescue. J. B. S. Haldane, Peter Medawar and George Williams separately put together the most satisfying account of the ageing process. Each species, it seems, comes equipped with a program of planned obsolescence chosen to suit its expected life-span and the age at which it is likely to have finished breeding. Natural selection carefully weeds out all genes that might allow damage to the body before or during reproduction. It does so by killing or lowering the reproductive success of all individuals that express such genes in youth. All the rest reproduce. But natural selection cannot weed out genes that damage the body in post-reproductive old age, because there is no reproduction of the successful in old age. Take Dunnet's fulmar, for instance. The reason it lives far longer than a mouse is because in the life of the fulmar there is no equivalent of the cat and the owl: no natural predators. A mouse is unlikely to make it past three years of age, so genes that damage four-year-old mouse bodies are under virtually no selection to die out. Fulmars are very likely to be around to breed at twenty, so genes that damage twenty-year-old fulmar bodies are still being ruthlessly weeded out.

Evidence for this theory comes from a natural experiment studied by Steven Austad on an island called Sapelo, which lies about five miles off the coast of Georgia in the United States. Sapelo contains a population of Virginia opossums that has been isolated for 10,000 years. Opossums, like many marsupials, age very rapidly. By the age of two years, opossums are generally dead from old age – the victims of cataracts, arthritis, bare skin and parasites. But that hardly matters because by two they have generally been hit by a truck, a coyote, an owl or some other natural enemy. On Sapelo, reasoned Austad, where many predators are absent, they would live longer and so – exposed for the first time to selection for better health after two years of age – their bodies would deteriorate less rapidly. They would age more slowly. This proved an accurate prediction. On Sapelo, Austad found, the opossums not only lived much longer, but aged more slowly. They were healthy enough to breed successfully in their second year – rare on the mainland – and their tendons showed less stiffness than those in mainland opossums.[8]

The evolutionary theory of ageing explains all the cross-species trends in a satisfying way. It explains why slow-ageing species tend to be large (elephants), or well protected (tortoises, porcupines), or relatively free from natural predators (bats, seabirds). In each case, because the death rate from accidents or predation is low, so the selective pressure is high for versions of genes that prolong health into later life.

Human beings, of course, have for several million years been large, well protected by weaponry (even chimps can chase leopards off with sticks) and have few natural predators. So we age slowly – and perhaps more slowly as the eras pass. Our infant mortality rate in a state of nature – of perhaps fifty per cent before the age of five – would be shockingly high by modern, western standards, but is actually low by the standards of other animals. Our Stone-Age ancestors began breeding at about twenty, continued until about thirty-five and looked after their children for about twenty years, so by about fifty-five they could die without damaging their reproductive success. Little wonder that at some time between fifty-five and seventy-five most of us gradually start to go grey, stiff, weak, creaky and deaf. All our systems begin to break down at once, as in the old story of the Detroit car maker who employed somebody to go around breakers' yards finding out which parts of cars did not break down, so that those bits could in future be made to a lower specification. Natural selection has designed all parts of our bodies to last just long enough to see our children into independence, no more.

Natural selection has built our telomeres of such a length that they can survive at most seventy-five to ninety years of wear, tear and repair. It is not yet known for certain, but it seems likely that natural selection may have given fulmars and tortoises somewhat longer telomeres, and Virginia opossums much shorter ones. Perhaps even the individual differences in longevity between one human being and another also indicate differences in telomere length. Certainly, there is great variety in telomere length between different people, from about 7,000 DNA 'letters' to about 10,000 per chromosome end. And telomere length is strongly inherited, as is longevity. People from long-lived families, in which members regularly reach ninety, may have longer telomeres, that take longer to fray, than the rest of us. Jeanne Calment, the French woman from Arles who in February 1995 became the first human being with a birth certificate to celebrate her 120th birthday, may have had many more repeats of the message TTAGGG. She eventually died at 122. Her brother lived to ninety-seven.[9]

In practice, though, it is more likely that Mme Calment could thank other genes for her longevity. Long telomeres are no good if the body decays rapidly; the telomeres will soon be shortened by the need for cell division to repair damaged tissues. In Werner's syndrome, an inherited misfortune characterised by premature and early ageing, the telomeres do indeed get shorter much more rapidly than in other people, but they start out the same size. The reason they get shorter is probably that the body lacks the capability to repair properly the corrosive damage done by so-called free radicals – atoms with unpaired electrons created by oxygen reactions in the body. Free oxygen is dangerous stuff, as any rusty piece of iron can testify. Our bodies, too, are continually 'rusting' from the effects of oxygen. Most of the mutations that cause 'longevity', at least in flies and worms, turn out to be in genes that inhibit the production of free radicals – i.e., they prevent the damage being done in the first place, rather than prolong the replicating life of cells that repair the damage. One gene, in nematode worms, has enabled scientists to breed a strain that lives to such an exceptional age that they would be 350 years old if they were human beings. In fruit flies, Michael

Rose has been selecting for longevity for twenty-two years: that is, in each generation he breeds from the flies that live the longest. His 'Methuselah' flies now live for 120 days, or twice as long as wild fruit flies, and start breeding at an age when wild fruit flies usually die. They show no sign of reaching a limit. A study of French centenarians quickly turned up three different versions of a gene on chromosome 6 that seemed to characterise long-lived people. Intriguingly, one of them was common in long-lived men and another was common in long-lived women.[10]

Ageing is turning out to be one of those things that is under the control of many genes. One expert estimates that there are 7,000 age-influencing genes in the human genome, or ten per cent of the total. This makes it absurd to speak of any gene as 'an ageing gene' let alone '*the* ageing gene'. Ageing is the more or less simultaneous deterioration of many different bodily systems; genes that determine the function of any of these systems can cause ageing, and there is good evolutionary logic in it. Almost any human gene can accumulate with impunity mutations which cause deterioration after breeding age.[11]

It is no accident that the immortal cell lines used by scientists in the laboratory are derived from cancer patients. The most famous of them, the HeLa cell line, originated in the cervical tumour of a patient named Henrietta Lacks, a black woman who died in Baltimore in 1951. Her cancer cells are so wildly proliferative when cultured in the laboratory that they often invade other laboratory samples and take over the Petri dish. They even somehow reached Russia in 1972 where they fooled scientists into thinking they had found new cancer viruses. HeLa cells were used for developing polio vaccines and have gone into space. Worldwide, they now weigh more than 400 times Henrietta's own body weight. They are spectacularly immortal. Yet nobody, at any time, thought to ask Henrietta Lacks's permission or that of her family – who were hurt when they learnt of her cellular immortality. In belated recognition of a 'scientific heroine', the city of Atlanta now recognises 11 October as Henrietta Lacks Day.

HeLa cells plainly have excellent telomerase. If antisense RNA is added to HeLa cells – that is, RNA containing the exact opposite message to the RNA message in telomerase, so that it will stick to the telomerase RNA – then the effect is to block the telomerase and prevent it working. The HeLa cells are then no longer immortal. They senesce and die after about twenty-five cell divisions.[12]

Cancer requires active telomerase. A tumour is invigorated with the biochemical elixir of youth and immortality. Yet cancer is the quintessential disease of ageing. Cancer rates rise steadily with age, more rapidly in some species than others, but still they rise: there is no creature on earth that is less likely to get cancer in old age than in youth. The prime risk factor for cancer is age. Environmental risk factors, such as cigarette smoking, work in part because they accelerate the ageing process: they damage the lungs, which require repair and repair uses up telomere length, thus making the cells 'older' in telomere terms than they would otherwise be. Tissues that are especially prone to cancer tend to be tissues that do a lot of cell division throughout life either for repair or for other reasons: skin, testis, breast, colon, stomach, white blood cells.

So we have a paradox. Shortened telomeres mean higher cancer risk, but telomerase, which keeps telomeres long, is necessary for a tumour. The resolution lies in the fact that the switching on of telomerase is one of the essential mutations that must occur if a cancer is to turn malignant. It is now fairly obvious why Geron's cloning of the telomerase gene caused its share price to rocket on the hopes of a general cure for cancer. Defeating telomerase would condemn tumours to suffer from the rapid advance of old age themselves.

Sex

All women become like their mothers. That is their
tragedy. No man does. That's his.
    Oscar Wilde, The Importance of Being Earnest

In the Prado Museum in Madrid hangs a pair of paintings by the
seventeenth-century court painter Juan Carreño de Miranda, called
'La Monstrua vestida' and 'La Monstrua desnuda': the monster
clothed and the monster naked. They show a grossly fat but very
unmonstrous five-year-old girl called Eugenia Martinez Vallejo.
There is indeed clearly something wrong with her: she is obese,
enormous for her age, has tiny hands and feet and strange-shaped
eyes and mouth. She was probably exhibited as a freak at a circus.
With hindsight, it is plain that she shows all the classic signs of a rare
inherited disease called Prader–Willi syndrome, in which children are
born floppy and pale-skinned, refuse to suck at the breast but later
eat till they almost burst, never apparently experiencing satiety, and
so become obese. In one case, the parent of a Prader–Willi child
found the child had consumed a pound of raw bacon in the back
of a car while being driven back from the shop. People with this

syndrome have small hands and feet, underdeveloped sex organs and they are also mildly mentally retarded. At times they throw spectacular temper tantrums, especially when refused food, but they also show what one doctor calls 'exceptional proficiency with jigsaw puzzles'.[1]

Prader–Willi syndrome was first identified by Swiss doctors in 1956. It might have been just another rare genetic disease, of the kind I have repeatedly promised not to write about in this book because GENES ARE NOT THERE TO CAUSE DIS-EASES. But there is something very odd about this particular gene. In the 1980s doctors noticed that Prader–Willi syndrome sometimes occurs in the same families as a completely different disease, a disease so different it might almost be called the opposite of Prader–Willi: Angelman's syndrome.

Harry Angelman was a doctor working in Warrington in Lancashire when he first realised that rare cases of what he called 'puppet children' were suffering from an inherited disease. In contrast to those with Prader–Willi syndrome, they are not floppy, but taut. They are thin, hyperactive, insomniac, small-headed and long-jawed, and often stick out their large tongues. They move jerkily, like puppets, but have a happy disposition; they are perpetually smiling and are given to frequent paroxysms of laughter. But they never learn to speak and are severely mentally retarded. Angelman children are much rarer than Prader–Willi children, but they sometimes crop up in the same family tree.[2]

In both Prader–Willi and Angelman's syndrome it soon became clear that the same chunk of chromosome 15 was missing. The difference was that in Prader–Willi syndrome, the missing chunk was from the father's chromosome, whereas in Angelman's syndrome, the missing chunk was from the mother's chromosome. Transmitted through a man, the disease manifests itself as Prader–Willi syndrome; transmitted through a woman it manifests itself as Angelman's syndrome.

These facts fly in the face of everything we have learnt about genes since Gregor Mendel. They seem to belie the digital nature of the genome and imply that a gene is not just a gene but carries with it some secret history of its origin. The gene 'remembers' which parent it came from because it is endowed at conception with a paternal or a maternal imprint – as if the gene from one parent were written in italic script. In every cell where the gene is active, the 'imprinted' version of the gene is switched on and the other version switched off. The body therefore expresses only the gene it inherited from the father (in the case of the Prader–Willi gene) or the mother (in the case of the Angelman gene). How this happens is still almost entirely obscure, though there is the beginning of an understanding. Why it happens is the subject of an extraordinary and daring evolutionary theory.

In the late 1980s, two groups of scientists, one in Philadelphia and one in Cambridge, made a surprising discovery. They tried to create a uniparental mouse – a mouse with only one parent. Since strict cloning from a body cell was then impossible in mice (post-Dolly, this is quickly changing), the Philadelphia team swapped the 'pronuclei' of two fertilised eggs. When an egg has been fertilised by a sperm, the sperm nucleus containing the chromosomes enters the egg but does not at first fuse with the egg nucleus: the two nuclei are known as 'pronuclei'. A clever scientist can sneak in with his pipette and suck out the sperm pronucleus, replacing it with the egg pronucleus from another egg – and vice versa. The result is two viable eggs, but one with, genetically speaking, two fathers and no mother and the other with two mothers and no father. The Cambridge team used a slightly different technique to reach the same result. But in both cases such embryos failed to develop properly and soon died in the womb.

In the two-mothers case, the embryo itself was properly organised, but it could not make a placenta with which to sustain itself. In the two-fathers case, the embryo grew a large and healthy placenta and most of the membranes that surround the foetus. But inside, where the embryo should be, there was a disorganised blob of cells with no discernible head.[3]

These results led to an extraordinary conclusion. Paternal genes, inherited from the father, are responsible for making the placenta; maternal genes, inherited from the mother, are responsible for making the greater part of the embryo, especially its head and brain. Why should this be? Five years later, David Haig, then at Oxford, thought he knew the answer. He had begun to reinterpret the mammalian placenta, not as a maternal organ designed to give sustenance to the foetus, but more as a foetal organ designed to parasitise the maternal blood supply and brook no opposition in the process. He noted that the placenta literally bores its way into the mother's vessels, forcing them to dilate, and then proceeds to produce hormones which raise the mother's blood pressure and blood sugar. The mother responds by raising her insulin levels to combat this invasion. Yet, if for some reason the foetal hormone is missing, the mother does not need to raise her insulin levels and a normal pregnancy ensues. In other words, although mother and foetus have a common purpose, they argue fiercely about the details of how much of the mother's resources the foetus may have – exactly as they later will during weaning.

But the foetus is built partly with maternal genes, so it would not be surprising if these genes found themselves with, as it were, a conflict of interest. The father's genes in the foetus have no such worries. They do not have the mother's interest at heart, except insofar as she provides a home for them. To turn briefly anthropomorphic, the father's genes do not trust the mother's genes to make a sufficiently invasive placenta; so they do the job themselves. Hence the paternal imprinting of placental genes as discovered by the two-fathered embryos.

Haig's hypothesis made some predictions, many of which were soon borne out. In particular, it predicted that imprinting would not occur in animals that lay eggs, because a cell inside an egg has no means of influencing the investment made by the mother in yolk size: it is outside the body before it can manipulate her. Likewise, even marsupials such as kangaroos, with pouches in place of placentas, would not, on Haig's hypothesis, have imprinted genes. So far, it appears, Haig is right. Imprinting is a feature of placental mammals and of plants whose seeds gain sustenance from the parent plant.[4]

Moreover, Haig was soon triumphantly noting that a newly discovered pair of imprinted genes in mice had turned up exactly where he expected them: in the control of embryonic growth. IGF2 is a miniature protein, made by a single gene, that resembles insulin. It is common in the developing foetus and switched off in the adult. IGF2R is a protein to which IGF2 attaches itself for a purpose that remains unclear. It is possible that IGF2R is there simply to get rid of IGF2. Lo and behold, both the *IGF2* and the *IGF2R* genes are imprinted: the first being expressed only from the paternal chromosome, the second from the maternal one. It looks very much like a little contest between the paternal genes trying to encourage the growth of the embryo and the maternal ones trying to moderate it.[5]

Haig's theory predicts that imprinted genes will generally be found in such antagonistic pairs. In some cases, even in human beings, this does seem to be the case. The human *IGF2* gene on chromosome 11 is paternally imprinted and when, by accident, somebody inherits two paternal copies, they suffer from Beckwith–Wiedemann syndrome, in which the heart and liver grow too large, and tumours of embryonic tissues are common. Although in human beings *IGF2R* is not imprinted, there does seem to be a maternally imprinted gene, *H19*, that opposes *IGF2*.

If imprinted genes exist only to combat each other, then you should be able to switch both off and it will have no effect at all on the development of the embryo. You can. Elimination of all imprinting leads to normal mice. We are back in the familiar territory of chromosome 8, where genes are selfish and do things for the benefit of themselves, not for the good of the whole organism. There is almost certainly nothing intrinsically purposeful about imprinting (though many scientists have speculated otherwise); it is another illustration of the theory of the selfish gene and of sexual antagonism in particular.

Once you start thinking in selfish-gene terms, some truly devious ideas pop into your head. Try this one. Embryos under the influence of paternal genes might behave differently if they share the womb with full siblings or if they share the womb with embryos that have different fathers. They might have more selfish paternal genes in the latter case. Having thought the thought, it was comparatively easy to do the deed and test this prediction with a natural experiment. Not all mice are equal. In some species of mice, for example *Peromyscus maniculatus*, the females are promiscuous, and each litter generally contains babies fathered by several different males. In other species, for example *Peromyscus polionatus*, the females are strictly monogamous and each litter contains full siblings who share both father and mother.

So what happens when you cross a *P. maniculatus* mouse with a *P. polionatus* mouse? It depends on which species is the father and which is the mother. If the promiscuous *P. maniculatus* is the father, the babies are born giant-sized. If the monogamous *P. polionatus* is the father, the babies are born tiny. Do you see what is happening? Paternal *maniculatus* genes, expecting to find themselves in a womb with competitors that are not even related, have been selected to fight for their share of the mother's resources at the expense of their co-foetuses. Maternal *maniculatus* genes, expecting to find embryos in their wombs that fight hard for her resources, have been selected to fight back. In the more neutral environment of *polionatus* wombs, the aggressive *maniculatus* genes from the father encounter only token opposition, so they win their particular battle: the baby is big if fathered by the promiscuous father and small if mothered by the promiscuous mother. It is a very neat demonstration of the imprinting theory.[6]

Neat as this tale is, it cannot be told without a caveat. Like many of the most appealing theories it may be too good to be true. In particular, it makes a prediction that is not borne out: that imprinted genes will be relatively rapidly evolving ones. This is because sexual antagonism would drive a molecular arms race in which each benefited from temporarily gaining the upper hand. A species-by-species comparison of imprinted genes does not bear this out. Rather, imprinted genes seem to evolve quite slowly. It looks increasingly as if the Haig theory explains some, but not all, cases of imprinting.[7]

Imprinting has a curious consequence. In a man, the maternal copy of chromosome 15 carries a mark that identifies it as coming from his mother, but when he passes it on to his son or daughter, it must somehow have acquired a mark that identifies it as coming from him: the father. It must switch from maternal to paternal and vice versa in the mother. That this switch does happen we know, because in a small proportion of people with Angelman syndrome there is nothing unusual about either chromosome except that both behave as if they were paternal. These are cases in which the switch failed to occur. They can be traced back to mutations in the previous generation, mutations that affect something called the imprinting centre, a small stretch of DNA close to both relevant genes, which somehow places the parental mark on the chromosome. The mark consists of one gene's methylation, of the kind encountered in chromosome 8.[8]

Methylation of the 'letter' C, you will recall, is the means by which genes are silenced, and it serves to keep selfish DNA under house arrest. But methylation is removed during the early development of the embryo – the creation of the so-called blastocyst – and then reimposed during the next stage of development, called gastrulation. Somehow, imprinted genes escape this process. They resist the demethylation. There are intriguing hints about how this is achieved, but nothing definitive.[9]

That imprinted genes escape demethylation is, we now know, all that stood between science and the cloning of mammals for many years. Toads were fairly easily cloned by putting genes from a body cell into a fertilised egg, but it just didn't work with mammals, because the genome of a female's body cells had certain critical genes switched off by methylation and the genome of a male's body cells had other genes switched off – the imprinted genes. So, confidently, scientists followed the discovery of imprinting with the announcement that cloning a mammal was impossible. A cloned mammal would be born with all its imprinted genes either on or off on both chromosomes, upsetting the doses required by the cells of the animal and causing development to fail. 'A logical consequence', wrote the scientists who discovered imprinting,[10] 'is the unlikelihood of successful cloning of mammals using somatic cell nuclei.'

Then, suddenly, along came Dolly the cloned Scottish sheep in early 1997. Quite how she and those that came after evaded the imprinting problem remains a mystery, even to her creators, but it seems that a certain part of the treatment meted out to her cells during the procedure erased all genetic imprints.[11]

The imprinted region of chromosome 15 contains about eight genes. One of these is responsible, when broken, for Angelman syndrome: a gene called *UBE3A*. Immediately beside this gene are two genes that are candidates for causing Prader–Willi syndrome when broken, one called *SNRPN* and the other called *IPW*. There could be others, but let us assume for the moment that *SNRPN* is the culprit.

The diseases, though, do not always result from a mutation in one of these genes but from an accident of a different kind. When an egg is formed inside a woman's ovary, it usually receives one copy of each chromosome, but in rare cases where a pair of parental chromosomes fails to separate, the egg ends up with two copies. After fertilisation with a sperm, the embryo now has three copies of that chromosome, two from the mother and one from the father. This is especially likely in elder mothers, and is generally fatal to the egg. The embryo can go on to develop into a viable foetus and survive more than a few days after birth only if the triplicate chromosome is number 21, the smallest of the chromosomes – the result being Down syndrome. In other cases the extra chromosome would so upset the biochemistry of the cells that development would fail.

However, in most cases, before that stage is reached, the body has a way of dealing with this triplet problem. It 'deletes' one chromosome altogether, leaving two, as intended. The difficulty is that it does so at random. It cannot be sure that it is deleting one of the two maternal chromosomes, or the single paternal one. Random deletion has a sixty-six per cent chance of getting one of the maternal ones, but accidents do happen. If, by mistake, it deletes the paternal one, then the embryo goes merrily on its way with two maternal chromosomes. In most cases this could not matter less, but if the tripled chromosome is number 15, you can see immediately what will ensue. Two copies of *UBE3A*, the maternally imprinted gene, are expressed, and no copies of *SNRPN*, the paternally imprinted gene. The result is Prader–Willi syndrome.[12]

Superficially, *UBE3A* does not look a very interesting gene. Its protein product is a type of 'E3 ubiquitin ligase', members of an obscure proteinaceous middle management within certain skin and lymph cells. Then in the middle of 1997, three different groups of scientists suddenly discovered that, in both mice and human beings, *UBE3A* is switched on in the brain. This is dynamite. The symptoms of both Prader–Willi and Angelman indicate something unusual about the brains of their victims. What is even more striking is that there is good evidence that other imprinted genes are active in the brain. In particular, it seems that in mice much of the forebrain is built by maternally imprinted genes, while much of the hypothalamus, at the base of the brain, is built by paternally imprinted genes.[13]

This imbalance was discovered by an ingenious piece of scientific work: the creation of mouse 'chimeras'. Chimeras are fused bodies of two genetically distinct individuals. They occur naturally – you may have met some or even be one yourself, though you will not know it without a detailed study of the chromosomes. Two genetically distinct embryos happen to fuse together and grow as if they were one. Think of them as the opposite of identical twins: two different genomes in one body, instead of two different bodies with the same genome.

It is comparatively easy to make mouse chimeras in the laboratory by gently fusing the cells from two early embryos. But what the ingenious Cambridge team did in this case was to fuse a normal mouse embryo with an embryo that was made by 'fertilising' an egg with another egg's nucleus, so that it had purely maternal genes and no contribution from the father. The result was a mouse with an unusually large head. When these scientists made a chimera between a normal embryo and an embryo derived only from the father (i.e., grown from an egg whose nucleus had been replaced by two sperm nuclei), the result was the opposite: a mouse with a big body and a small head. By equipping the maternal cells with the biochemical equivalent of special radio transmitters to send out signals of their presence, they were able to make the remarkable discovery that most of the striatum, cortex and hippocampus of the mouse brain are consistently made by these maternal cells, but that such cells are excluded from the hypothalamus. The cortex is the place where sensory information is processed and behaviour is produced. Paternal cells, by contrast, are comparatively scarce in the brain, but much commoner in the muscles. Where they do appear in the brain, however, they contribute to the development of the hypothalamus, amygdala and preoptic area. These areas comprise part of the 'limbic system' and are responsible for the control of emotions. In the opinion of one scientist, Robert Trivers, this difference reflects the fact that the cortex has the job of co-operating with maternal relatives while the hypothalamus is an egotistical organ.[14]

In other words, if we are to believe that the placenta is an organ that the father's genes do not trust the mother's genes to make, then the cerebral cortex is an organ that the mother's genes do not trust the father's genes to make. If we are like mice, we may be walking around with our mothers' thinking and our fathers' moods (to the extent that thoughts and moods are inherited at all). In 1998 another imprinted gene came to light in mice, which had the remarkable property of determining a female mouse's maternal behaviour. Mice with this Mest gene intact are good, caring mothers to their pups. Female mice who lack a working copy of the gene are also normal except that they make terrible mothers. They fail to build decent nests, they fail to haul their pups back to the nest when they wander, they do not keep the pups clean and they generally seem not to care. Their pups usually die. Inexplicably, the gene is paternally inherited. Only the version inherited from the father functions; the mother's version remains silent.[15]

The Haig theory of conflict over embryonic growth does not easily explain these facts. But the Japanese biologist Yoh Iwasa has a theory that does. He argues that because the father's sex chromosome determines the sex of the offspring – if he passes on an X rather than a Y chromosome, the offspring is female – so paternal X chromosomes are found only in females. Therefore, behaviour that is characteristically required of females should be expressed only from paternal chromosomes. If they were also expressed from maternal X chromosomes, they might appear in males, or they might be overexpressed in females. It therefore makes sense that maternal behaviour should be paternally imprinted.[16]

The best vindication of this idea comes from an unusual natural experiment studied by David Skuse and his colleagues at the Institute of Child Health in London. Skuse located eighty women and girls aged between six and twenty-five who suffered from Turner's syndrome, a disorder caused by the absence of all or part of the X chromosome. Men have only one X chromosome, and women keep one of their two X chromosomes switched off in all their cells, so Turner's syndrome should, in principle, make little difference to development. Indeed, Turner's girls are of normal intelligence and appearance. However, they often have trouble with 'social adjustment'. Skuse and his colleagues decided to compare two kinds of Turner's girls: those with the paternal X chromosome missing and those with the maternal X chromosome missing. The twenty-five girls missing the maternal chromosome were significantly better adjusted, with 'superior verbal and higher-order executive function skills, which mediate social interactions' than the fifty-five girls missing the paternal chromosome. Skuse and his colleagues determined this by setting the children standard tests for cognition, and giving the parents questionnaires to assess social adjustment. The questionnaire asked the parents if the child lacked awareness of other people's feelings, did not realise when others were upset or angry, was oblivious to the effect of her behaviour on other members of the family, was very demanding of people's time, was difficult to reason with when upset, unknowingly offended people with her behaviour, did not respond to commands, and other similar questions. The parents had to respond with 0 (for 'not at all true'), 1 for 'quite or sometimes true' and 2 for 'very or often true'. The total from all twelve questions was then totted up. All the Turner's girls had higher scores than normal girls and boys, but the ones who were lacking the paternal X chromosome had more than twice the score of the ones lacking the maternal X chromosome.

The inference is that there is an imprinted gene somewhere on the X chromosome, which is normally switched on only in the paternal copy and that this gene somehow enhances the development of social adjustment – the ability to understand other's feelings, for example. Skuse and his colleagues provided further evidence of this from children who were missing only part of one X chromosome.[17]

This study has two massive implications. First, it suggests an explanation for the fact that autism, dyslexia, language impairment and other social problems are much commoner in boys than girls. A boy receives only one X chromosome, from his mother, so he presumably gets one with the maternal imprint and the gene in question switched off. As of this writing, the gene has not been located, but imprinted genes are known from the X chromosome.

But second, and more generally, we are beginning to glimpse an end to the somewhat ridiculous argument over gender differences that has continued throughout the late twentieth century and has pitted nature against nurture. Those in favour of nurture have tried to deny any role for nature, while those who favour nature have rarely denied a role for nurture. The question is not whether nurture has a role to play, because nobody of any sense has ever gone on record as denying that it does, but whether nature has a role to play at all. When my one-year-old daughter discovered a plastic baby in a toy pram one day while I was writing this chapter, she let out the kinds of delighted squeals that her brother had reserved at the same age for passing tractors. Like many parents, I found it hard to believe that this was purely because of some unconscious social conditioning that we had imposed. Boys and girls have systematically different interests from the very beginning of autonomous behaviour. Boys are more competitive, more interested in machines, weapons and deeds. Girls are more interested in people, clothes and words. To put it more boldly, it is no thanks only to upbringing that men like maps and women like novels.

In any case, the perfect, if unconscionably cruel, experiment has been done by the supporters of pure nurture. In the 1960s, in the United States, a botched circumcision left a boy with a badly damaged penis, which the doctor decided to amputate. It was decided to try to turn the boy into a girl by castration, surgery and hormonal treatment. John became Joan; she wore dresses and played with dolls. She grew up into a young woman. In 1973 John Money, a Freudian psychologist, claimed in a burst of publicity that Joan was a well adjusted adolescent, and her case thus put an end to all speculation: gender roles were socially constructed.

Not until 1997 did anybody check the facts. When Milton Diamond and Keith Sigmundson tracked down Joan, they found a man, happily married to a woman. His story was very different from that told by Money. He had always felt deeply unhappy about something as a child and had always wanted to wear trousers, mix with boys and urinate standing up. At the age of fourteen he was told by his parents what had happened, which brought great relief. He ceased hormonal treatment, changed his name back to John, resumed the life of a man, had his breasts removed and at the age of twenty-five married a woman and adopted her children. Held up as a proof of socially constructed gender roles, he proved the exact opposite: that nature does play a role in gender. The evidence from zoology has always pointed that way: male behaviour is systematically different from female behaviour in most species and the difference has an innate component. The brain is an organ with innate gender. The evidence from the genome, from imprinted genes and genes for sex-linked behaviours, now points to the same conclusion.[18]

Memory

Heredity provides for the modification of its own
machinery.   *James Mark Baldwin, 1896*

The human genome is a book. By reading it carefully from beginning
to end, taking due account of anomalies like imprinting, a skilful
technician could make a complete human body. Given the right
mechanism for reading and interpreting the book, an accomplished
modern Frankenstein could carry out the feat. But what then? He
would have made a human body and imbued it with the elixir of
life, but for it to be truly alive it would have to do more than exist.
It would have to adapt, to change and to respond. It would have
to gain its autonomy. It would have to escape Frankenstein's control.
There is a sense in which the genes, like the hapless medical student
in Mary Shelley's story, must lose control of their own creation.
They must set it free to find its own path through life. The genome
does not tell the heart when to beat, nor the eye when to blink, nor
the mind when to think. Even if the genes do set some of the
parameters of personality, intelligence and human nature with surprising precision, they know when to delegate. Here on chromosome

16 lie some of the great delegators: genes that allow learning and memory.

We human beings may be determined to a surprising extent by the dictates of our genes, but we are determined even more by what we have learnt in our lifetimes. The genome is an information-processing computer that extracts useful information from the world by natural selection and embodies that information in its designs. Evolution is just terribly slow at processing the information, needing several generations for every change. Little wonder that the genome has found it helpful to invent a much faster machine, whose job is to extract information from the world in a matter of minutes or seconds and embody that information in behaviour – the brain. Your genome supplies you with the nerves to tell when your hand is hot. Your brain supplies you with the action to remove your hand from the stove-top.

The subject of learning lies in the provinces of neurosciences and psychology. It is the opposite of instinct. Instinct is genetically-determined behaviour; learning is behaviour modified by experience. The two have little in common, or so the behaviourist school of psychology would have had us all believe during much of the twentieth century. But why are some things learnt and others instinctive? Why is language an instinct, while dialect and vocabulary are learnt? James Mark Baldwin, the hero of this chapter, was an obscure American evolutionary theorist of the last century, who wrote an article in 1896 summarising a dense and philosophical argument that had little influence at the time, or indeed at any time during the subsequent ninety-one years. But by a stroke of good fortune, he was plucked from obscurity by a group of computer scientists in the late 1980s, who decided his argument was of great relevance to their problem of teaching computers how to learn.[1]

What Baldwin wrestled with was the question of why something is learnt by an individual in his lifetime rather than pre-programmed as an instinct. There is a commonly held belief that learning is good, instinct bad – or, rather, that learning is advanced and instinct primitive. It is therefore a mark of human rank that we need to learn all sorts of things that come naturally to animals. Artificial-Intelligence researchers, following this tradition, quickly placed learning on a pinnacle: their goal was the general-purpose learning machine. But this is just a factual mistake. Human beings achieve by instinct the same things that animals do. We crawl, stand, walk, cry and blink in just as instinctive a way as a chick. We employ learning only for the extra things we have grafted on to the animal instincts: things like reading, driving, banking and shopping. 'The main function of consciousness', wrote Baldwin, 'is to enable [the child] to learn things which natural heredity fails to transmit.'

And by forcing ourselves to learn something, we place ourselves in a selective environment that puts a premium on a future instinctive solution to the problem. Thus, learning gradually gives way to instinct. In just the same way, as I suggested in the chapter on chromosome 13, the invention of dairy farming presented the body with the problem of the indigestibility of lactose. The first solution was cultural – to make cheese – but later the body evolved an innate solution by retaining lactase production into adulthood. Perhaps even literacy would become innate eventually if illiterate people were at a reproductive disadvantage for long enough. In effect, since the process of natural selection is one of extracting useful information from the environment and encoding it in the genes, there is a sense in which you can look on the human genome as four billion years' worth of accumulated learning.

However, there comes a limit to the advantages of making things innate. In the case of spoken language, where we have a strong instinct, but a flexible one, it would clearly be madness for natural selection to go the whole hog and make even the vocabulary of the language instinctive. That way language would have been far too inflexible a tool: lacking a word for computer, we would have to describe it as 'the thing that thinks when you communicate with it'. Likewise, natural selection has taken care (forgive the teleological shorthand) to equip migratory birds with a star-navigation system that is not fully assembled. Because of the precession of the equinoxes, which gradually changes the direction of North, it is vital that birds recalibrate their star compass in every generation through learning.

The Baldwin effect is about the delicate balance between cultural and genetic evolution. They are not opposites, but comrades, trading influence with each other to get best results. An eagle can afford to learn its trade from its parents the better to adapt to local conditions; a cuckoo, by contrast, must build everything into instinct because it will never meet its parents. It must expel its foster siblings from the nest within hours of hatching; migrate to the right part of Africa in its youth with no parents to guide it; discover how to find and eat caterpillars; return to its birthplace the following spring; acquire a mate; locate the nest of a suitable host bird – all by a series of instinctive behaviours with judicious bouts of learning from experience.

Just as we underestimate the degree to which human brains rely upon instincts, so we have generally underestimated the degree to which other animals are capable of learning. Bumble bees, for instance, have been shown to learn a great deal from experience about how to gather nectar from different types of flowers. Trained on one kind, they are incompetent at another until they have had practice; but once they know how to deal with, say, monkshood, they are also better at dealing with similar-shaped flowers such as lousewort – thus proving that they have done more than memorise individual flowers, but have generalised some abstract principles.

Another famous example of animal learning in an equally simple creature is the case of the sea slug. A more contemptibly basic animal is hard to imagine. It is slothful, small, simple and silent. It has a minute brain and it lives its life of eating and sex with an enviable lack of neurosis. It cannot migrate, communicate, fly or think. It just exists. Compared with, say, a cuckoo or even a bumble bee, its life is a doddle. If the idea that simple animals use instincts and complicated ones learn is right, then the sea slug has no need of learning.

Yet learn it can. If a jet of water is blown upon its gill, it withdraws the gill. But if the jet of water is repeatedly blown on the gill, the withdrawal gradually ceases. The sea slug stops responding to what it now recognises as a false alarm. It 'habituates'. This is hardly learning the differential calculus, but it is learning all the same. Conversely, if given an electric shock once, before water is blown on the gill, the sea slug learns to withdraw its gill even further than usual – a phenomenon called sensitisation. It can also be 'classically conditioned', like Pavlov's famous dogs, to withdraw its gill when it receives only a very gentle puff of water if that gentle puff is paired with an electric shock: thereafter, the gentle puff alone, normally insufficient to make the sea slug withdraw its gill, results in a rapid gill withdrawal. Sea slugs, in other words, are capable of the same kinds of learning as dogs or people: habituation, sensitisation and associative learning. Yet they do not even use their brains. These reflexes and the learning that modifies them occur in the abdominal ganglion, a small nervous substation in the belly of the slimy creature.

The man behind these experiments, Eric Kandel, had a motive other than bothering slugs. He wanted to understand the basic mechanism by which learning occurred. What is learning? What changes occur to nerve cells when the brain (or the abdominal ganglion) acquires a new habit or a change in its behaviour? The central nervous system consists of lots of nerve cells, down each of which electrical signals travel; and synapses, which are junctions between nerve cells. When an electrical nerve signal reaches a synapse, it must transfer to a chemical signal, like a train passenger catching a ferry across a sea channel, before resuming its electrical journey. Kandel's attention quickly focused on these synapses between neurons. Learning seems to be a change in their properties. Thus when a sea slug habituates to a false alarm, the synapse between the receiving, sensory neuron and the neuron that moves the gill is somehow weakened. Conversely, when the sea slug is sensitised to the stimulus, the synapse is strengthened. Gradually and ingeniously, Kandel and his colleagues homed in on a particular molecule in the sea-slug brain which lay at the heart of this weakening or strengthening of the synapses. The molecule is called cyclic AMP.

Kandel and his colleagues discovered a cascade of chemical

changes all centred around cyclic AMP. Ignoring their names, imagine a string of chemicals called A, B, C and so on:

A makes B,
Which activates C,
Which opens a channel called D,
Thus allowing more of E into the cell,
Which prolongs the release of F,
Which is the neurotransmitter that ferries the signal across the synapse to the next neuron.

Now it so happens that C also activates a protein called CREB by changing its form. Animals that lack this activated form of CREB can still learn things, but cannot remember them for more than an hour or so. This is because CREB, once activated, starts switching on genes and thus altering the very shape and function of the synapse. The genes thus alerted are called *CRE* genes, which stands for cyclic-AMP response elements. If I go into more detail I will drive you back to the nearest thriller, but bear with me, it is about to get simpler again.[2]

So simple, in fact, that it is time to meet *dunce*. *Dunce* is a mutant fruit fly incapable of learning that a certain smell is always followed by an electric shock. Discovered in the 1970s, it was the first of a string of 'learning mutants' to be discovered by giving irradiated flies simple tasks to learn and breeding from those that could not manage the tasks. Other mutants called *cabbage*, *amnesiac*, *rutabaga*, *radish* and *turnip* soon followed. (Once again, fruit-fly geneticists are allowed much more liberty with gene names than their human-genetics colleagues.) In all, seventeen learning mutations have now been found in flies. Alerted by the feats of Kandel's sea slugs, Tim Tully of Cold Spring Harbor Laboratory set out to find out exactly what was wrong with these mutant flies. To Tully's and Kandel's delight, the genes that were 'broken' in these mutants were all involved in making or responding to cyclic AMP.[3]

Tully then reasoned that if he could knock out the flies' ability to learn, he could alter or enhance it as well. By removing the gene for the CREB protein, he created a fly that could learn, but not remember that it had learnt – the lesson soon faded from its memory. And he soon developed a strain of fly that learnt so fast that it got the message after a single lesson whereas other flies needed ten lessons to learn to fear a smell that was reliably followed by an electric shock. Tully described these flies as having photographic memories; far from being clever, they over-generalised horribly, like a person who reads too much into the fact that the sun was shining when he had a bicycle accident and refuses thereafter to bicycle on sunny days. (Great human mnemonists, such as the famous Russian Sherashevsky, experience exactly this problem. They cram their heads with so much trivia that they cannot see the wood for the trees. Intelligence requires a judicious mixture of remembering and forgetting. I am often struck by the fact that I easily 'remember' – i.e., recognise – that I have read a particular piece of text before, or heard a particular radio programme, yet I could not have recited either: the memory is somehow hidden from my consciousness. Presumably, it is not so hidden in mnemonists' minds.)[4]

Tully believes that CREB lies at the heart of learning and memory mechanisms, a sort of master gene that switches on other genes. So the quest to understand learning becomes a genetic quest after all. Far from escaping from the tyranny of genes by discovering how to learn instead of behave instinctively, we have merely found that the surest way to understand learning is to understand the genes and their products that enable learning to occur.

By now, it will come as no surprise to learn that CREB is not confined to flies and slugs. Virtually the same gene is present in mice, as well, and mutant mice have already been created by knocking out the mouse *CREB* gene. As predicted they are incapable of simple learning tasks, such as remembering where the hidden under-water platform lies in a swimming bath (this is standard torture in mouse learning experiments) or remembering which foods were safe to eat. Mice can be made temporarily amnesiac by injecting the 'antisense', or opposite, of the *CREB* gene into their brains – this silences the gene for a while. Likewise, they are super-learners if their *CREB* gene is especially active.[5]

And from mice to men is but an evolutionary hair's breadth. We human beings have *CREB* genes, too. The human *CREB* gene itself is on chromosome 2, but its crucial ally, which helps *CREB* to do its job, called *CREBBP*, is right here on chromosome 16. Together with another 'learning' gene called alpha-integrin, also on chromosome 16, it provides me with a (somewhat weak) excuse for a chapter on learning.

In fruit flies the cyclic AMP system seems to be especially active in brain regions called mushroom bodies, toadstool-shaped extrusions of neurons in the fruit fly brain. If a fly has no mushroom bodies in its brain, then it is generally incapable of learning the association between a smell and an electric shock. CREB and cyclic AMP seem to do their work in these mushroom bodies. Exactly how is only now becoming clear. By systematically searching for other mutant flies incapable of learning or memory, Ronald Davis, Michael Grotewiel and their colleagues in Houston came up with a different kind of mutant fly, which they called *volado*. ('Volado', they helpfully explain, is a Chilean colloquialism meaning something akin to 'absent-minded' or 'forgetful', and generally applied to professors.) Like *dunce*, *cabbage* and *rutabaga*, *volado* flies have a hard time learning. But unlike those genes, *volado* seems to have nothing to do with CREB or cyclic AMP. It is the recipe for a subunit of a protein called an alpha-integrin, which is expressed in mushroom bodies, and which seems to play a role in binding cells together.

To check that this was not a 'chopstick' gene (see the chapter on chromosome 11) that had lots of effects beside altering memory, the Houston scientists did something rather clever. They took some flies in which the *volado* gene had been knocked out, and inserted a fresh copy linked with a 'heat-shock' gene – a gene that becomes switched on when suddenly heated up. They had carefully arranged the two so that the *volado* gene only worked when the heat-shock gene was on. At cool temperatures, the flies could not learn. Three hours after a heat shock, however, they suddenly became good learners. A few hours after that, as the heat shock faded into the past, they again lost the ability to learn. This means that *volado* is needed at the exact moment of learning; it is not just a gene required to build the structures that do the learning.[6]

The fact that the *volado* gene's job is to make a protein that binds cells together raises the intriguing hint that memory may consist, quite literally, of the tightening of the connections between neurons. When you learn something, you alter the physical network of your brain so as to create new, tight connections where there were none or weaker ones before. I can just about accept that this is what learning and memory consist of, but I have a hard time imagining how my memory of the meaning of the word 'volado' consists of some strengthened synaptic connections between a few neurons. It is distinctly mind-boggling. Yet far from having removed the mystery from the problem by reducing it to the molecular level, I feel that scientists have opened before me a new and intriguing mystery, the mystery of trying to imagine how connections between nerve cells not only provide the mechanism of memory but *are* memory. It is every bit as thrilling a mystery as quantum physics, and a great deal more thrilling than Ouija boards and flying saucers.

Let us delve a little deeper into the mystery. The discovery of *volado* hints at the hypothesis that integrins are central to learning and memory, but there were already hints of this kind. By 1990 it was already known that a drug that inhibited integrins could affect memory. Specifically, such a drug interfered with a process called long-term potentiation (LTP), which seems to be a key event in the creation of a memory. Deep in the base of the brain lies a structure called the hippocampus (Greek for sea-horse) and a part of the hippocampus is called the Ammon's horn (after the Egyptian god associated with the ram and later adopted as his 'father' by Alexander the Great after his mysterious visit to the Siwah oasis in Libya). In the Ammon's horn, in particular, there are a large number of 'pyramidal' neurons (note the continuing Egyptian theme) which gather together the inputs of other, sensory neurons. A pyramidal neuron is difficult to 'fire', but if two separate inputs arrive at once, their combined effect will fire it. Once fired, it is much easier to fire but only by one of the two inputs that originally fired it, and not by another input. Thus, the sight of a pyramid and the sound of the word 'Egypt' can combine to fire a pyramidal cell, creating an associative memory between the two, but the thought of a sea-horse, although perhaps connected to the same pyramidal cell, is not 'potentiated' in the same way because it did not coincide in time. That is an example of long-term potentiation. If you think, too simplistically, of the pyramidal cell as the memory of Egypt, then it can now be fired by the word or the picture, but not by a sea-horse.

Long-term potentiation, like sea-slug learning, absolutely depends on a change in the properties of synapses, in this case the synapses between the inputting cells and the pyramidal cells. That change almost certainly involves integrins. Oddly, the inhibition of integrins does not interfere with the formation of long-term potentiation, but it does interfere with its maintenance. Integrins are probably needed for literally holding the synapse closely together.

I glibly implied a few moments ago that the pyramidal cell might actually be a memory. This is nonsense. The memories of your childhood do not even reside in the hippocampus, but in the neo-cortex. What resides in and near the hippocampus is the mechanism for creating a new long-term memory. Presumably, the pyramidal cells in some manner transmit that newly formed memory to where it will reside. We know this because of two remarkable and unfortunate young men, who suffered bizarre accidents in the 1950s. The first, known in the scientific literature by the initials H.M., had a chunk of his brain removed to prevent the epileptic seizures caused by a bicycle accident. The second, known as N.A., was a radar technician in the air force, who one day was sitting building a model when he happened to turn round. A colleague, who was playing with a mini-ature fencing foil, chose that moment to stab forward and the foil passed through N.A.'s nostril and into his brain.

Both men suffer to this day from terrible amnesia. They can remember events from their childhood quite clearly and from right up to a few years before their accidents. They can memorise current events briefly if not interrupted before being asked to recall them. But they cannot form new long-term memories. They cannot recognise the face of somebody they see every day or learn their way home. In N.A.'s (milder) case, he cannot enjoy television because commercials cause him to forget what went before them.

H.M. can learn a new task quite well and retains the skill, but cannot recall that he has learnt it – which implies that procedural memories are formed somewhere different from 'declarative' memories for facts or events. This distinction is confirmed by a study of three young people with severe amnesia for facts and events, who were found to have gone through school, acquiring reading, writing and other skills with comparatively little difficulty. All three, on being scanned, proved to have unusually small hippocampuses.[7]

But we can get a little more specific than just saying that memories are made in hippocampuses. The damage that both H.M. and N.A. suffered implies a connection between two other parts of the brain and memory formation: the medial temporal lobe, which H.M. lacks, and the diencephalon, which N.A. partly lacks. Prompted by this, neuroscientists have gradually narrowed down the search for the most vital of all memory organs to one principal structure, the perirhinal cortex. It is here that sensory information, sent from the visual, auditory, olfactory or other areas, is processed and made into memories, perhaps with the help of CREB. The information is then passed to the hippocampus and thence to the diencephalon for temporary storage. If it is deemed worthy of permanent preservation it is sent back to the neo-cortex as a long-term memory: that strange moment when you suddenly don't need to keep looking up somebody's telephone number but can recall it. It seems probable that the transmission of memory from the medial temporal lobe to the neo-cortex happens at night during sleep: in rats' brains the cells of the lobe fire actively at night.

The human brain is a far more impressive machine than the genome. If you like quantitative measures, it has trillions of synapses instead of billions of bases and it weighs kilograms instead of micrograms. If you prefer geometry, it is an analogue, three-dimensional machine, rather than a digital, two-dimensional one. If you like thermodynamics, it generates large quantities of heat as it works, like a steam engine. For biochemists, it requires many thousands of different proteins, neurotransmitters and other chemicals, not just the four nucleotides of DNA. For the impatient, it literally changes while you watch, as synapses are altered to create learned memories, whereas the genome changes more slowly than a glacier. For the lover of free will, the pruning of the neural networks in our brains, by the ruthless gardener called experience, is vital to the proper functioning of the organ, whereas genomes play out their messages in a predetermined way with comparatively little flexibility. In every way, it seems, conscious, willed life has advantages over automatic, gene-determined life. Yet, as James Mark Baldwin realised and modern Artificial-Intelligence nerds appreciate, the dichotomy is a false one. The brain is created by genes. It is only as good as its innate design. The very fact that it is a machine designed to be modified by experience is written in the genes. The mystery of how is one of the great challenges of modern biology. But that the human brain is the finest monument to the capacities of genes there is no doubt. It is the mark of a great leader that he knows when to delegate. The genome knew when to delegate.

Death

If learning is making new connections between brain cells, it is also about losing old connections. The brain is born with far too many connections between cells; many are lost as it develops. For example, at first each side of the visual cortex is connected to one half of the input from both eyes. Only by fairly drastic pruning does this change so that one slice of the brain receives input from the right eye and another slice receives input from the left eye. Experience causes the unnecessary connections to wither away and thereby turns the brain from a general to a specific device. Like a sculptor chipping away at a block of marble to find the human form within, so the environment strips away the surplus neurons to sharpen the skills of the brain. In a blind, or permanently blindfolded young mammal, this sorting out never happens.

But the withering means more than the loss of synaptic connections. It also means the death of whole cells. A mouse with a faulty version of a gene called *ced-9* fails to develop properly because cells in its brain that are not needed fail to do their duty and die. The mouse ends up with a disorganised and overloaded brain that does not work. Folk wisdom loves to recite the grim (but meaningless) statistic that we lose a million brain cells a day. In our youth, and even in the womb, we do indeed lose brain cells at a rapid rate. If we did not, we would never be able to think at all.[1]

Prodded by genes like *ced-9*, the unneeded cells commit mass suicide (other *ced* genes cause suicide in other body tissues). The dying cells obediently follow a precise protocol. In microscopic nematode worms, the growing embryo eventually contains 1,090 cells, but precisely 131 of these kill themselves during development, leaving 959 cells in an adult worm. It is as if they sacrifice themselves for the greater good of the body. '*Dulce et decorum est pro corpore mori*' they cry and fade heroically away, like soldiers going over the top at Verdun, or worker bees suicidally stinging an intruder. The analogy is far from specious. The relationship between body cells is indeed very much like that between bees in a hive. The ancestors of your cells were once individual entities and their evolutionary 'decision' to co-operate, some 600 million years ago, is almost exactly equivalent to the same decision, taken perhaps fifty million years ago by the social insects, to co-operate on the level of the body: close genetic relatives discovered they could reproduce more effectively if they did so vicariously, delegating the task to germ cells in the cells' case, or to a queen in the case of bees.[2]

The analogy is so good that evolutionary biologists have begun to realise that the co-operative spirit goes only so far. Just as soldiers at Verdun were occasionally driven to mutiny against the greater good, so worker bees are capable of reproducing on their own if they get the chance; only the vigilance of other workers prevents them. The queen buys the loyalty of those other workers to herself rather than to their sister workers by mating with several males to ensure that most workers are only half-sisters of each other and therefore share little genetic common interest. And so it is with cells in the body. Mutiny is a perpetual problem. Cells are continually forgetting their patriotic duty, which is to serve the germ cells, and setting out to reproduce themselves. After all, each cell is descended from a long line of reproducing cells; it goes against the grain to cease dividing for a whole generation. And so, in every tissue every day, there is a cell that breaks ranks and starts to divide again, as if unable to resist the age-old call of the genes to reproduce themselves. If the cell cannot be stopped, we call the result cancer.

But usually, it can be stopped. The problem of cancerous mutiny is so old that in all large bodied animals the cells are equipped with an elaborate series of switches designed to induce the cell to commit suicide if it should find itself turning cancerous. The most famous and important of these switches, in fact possibly the most talked about of all human genes since its discovery in 1979, is *TP53*, which lies on the short arm of chromosome 17. This chapter tells the remarkable story of cancer, through the eyes of a gene whose principal job is to prevent it.

When Richard Nixon declared war on cancer in 1971, scientists did not even know what the enemy was, beyond the obvious fact that it was an excessive growth of tissue. Most cancer was plainly neither infectious nor inherited. The conventional wisdom was that cancer was not a single form of disease at all, but a collection of diverse disorders induced by a multiplicity of causes, most of them external. Chimney sweeps 'caught' scrotal cancer from coal tar; X-ray technicians and Hiroshima survivors contracted leukaemia from radiation; smokers 'caught' lung cancer from cigarette smoke and shipyard workers 'caught' the same affliction from asbestos fibres. There might be no common thread, but if there was it probably involved a failure of the immune system to suppress tumours. So went the conventional wisdom.

Two rival lines of research were, however, beginning to produce new insights that would lead to a revolution in the understanding of cancer. The first was the discovery in the 1960s by Bruce Ames in California that many chemicals and radiations that caused cancer, such as coal tar and X-rays, had one crucial thing in common: they were very good at damaging DNA. Ames glimpsed the possibility that cancer was a disease of the genes.

The second breakthrough had begun much earlier. In 1909, Peyton Rous had proved that a chicken with a form of cancer called sarcoma could pass the disease to a healthy chicken. His work was largely ignored, since there seemed so little evidence that cancer was contagious. But in the 1960s, a whole string of animal cancer viruses, or oncoviruses, were discovered, beginning with the Rous sarcoma virus itself. Rous was eventually given the Nobel prize at the age of eighty-six in recognition of his prescience. Human oncoviruses soon followed and it became apparent that whole classes of cancer, such as cervical cancer, were indeed caused partly by viral infection.[3]

Putting the Rous sarcoma virus through the gene-sequencer revealed that it carried a special cancer-causing gene, now known as *src*. Other such 'oncogenes' soon followed from other oncoviruses. Like Ames, the virologists were beginning to realise that cancer was a disease of genes. In 1975 the world of cancer research was turned upside down by the discovery that *src* was not a viral gene at all. It was a gene that we all possessed, chicken, mouse and human, too. The Rous sarcoma virus had stolen its oncogene from one of its hosts.

More conventional scientists were reluctant to accept that cancer was a genetic disease: after all, except in rare cases, cancer was not inherited. What they were forgetting was that genes are not confined to the germline; they also function during an organism's lifetime in every other organ. A genetic disease within an organ of the body, but not in the reproductive cells, could still be a genetic disease. By 1979, DNA taken from three kinds of tumour had been used to induce cancerous growth in mouse cells, thus proving that genes alone could cause cancer.

It was obvious from the start what kinds of genes oncogenes would turn out to be – genes that encourage cells to grow. Our cells possess such genes so that we can grow in the womb and in childhood, and so that we can heal wounds in later life. But it is vital that they are switched off most of the time; if they are jammed on, the result can be disastrous. With 100 trillion body cells, and a fairly rapid turnover, there are plenty of opportunities for oncogenes to be jammed on during a lifetime, even without the encouragement of mutation-causing cigarette smoke or sunlight. Fortunately, however, the body possesses genes whose job is to detect excessive growth and shut it down. These genes, discovered first in the mid-1980s by Henry Harris at Oxford, are known as tumour-suppressor genes. Tumour suppressors are the opposite of oncogenes. Whereas oncogenes cause cancer if they are jammed on, tumour-suppressor genes cause cancer if they are jammed off.

They do their job by various means, the most prominent of which is to arrest a cell at a certain point in its cycle of growth and division, then release it from arrest only if it has all its papers in order, so to speak. To progress beyond this stage, therefore, a tumour must contain a cell that has both a jammed-on oncogene and a jammed-off tumour-suppressor gene. That is unlikely enough, but it is not the end of the matter. To escape and grow uncontrollably, the tumour must now pass by an even more determined checkpoint, manned by a gene that detects abnormal behaviour in a cell and issues an instruction to different genes to dismantle the cell from the inside: to commit suicide. This is *TP53*.

When *TP53* was first discovered, by David Lane in Dundee in 1979, it was thought to be an oncogene, but it was later recognised to be a tumour suppressor. Lane and his colleague Peter Hall were discussing *TP53* in a pub one day in 1992 when Hall offered his arm as a guinea pig for testing if *TP53* was a tumour suppressor. Getting permission to perform an animal test would take months, but an experiment on a human volunteer could be done right away. Hall repeatedly scarred a small part of his arm with radiation and Lane took biopsies over the succeeding two weeks. They showed a dramatic rise in the level of p53, the protein manufactured from *TP53*, following the radiation damage, clear evidence that the gene responded to cancer-causing damage. Lane has gone on to develop p53 as a potential cancer cure in clinical trials; the first human volunteers will be taking the drug as this book is being published. Indeed, so rapidly has cancer research in Dundee grown that p53 is now bidding to be the third most famous product of the small Scottish city on the Tay estuary, after jute and marmalade.[4]

Mutation in the *TP53* gene is almost the defining feature of a lethal cancer; in fifty-five per cent of all human cancers, *TP53* is broken. The proportion rises to over ninety per cent among lung cancers. People born with one faulty version of *TP53* out of the two they inherit, have a ninety-five per cent chance of getting cancer, and usually at an early age. Take, as an example, colorectal cancer. This cancer begins with a mutation that breaks a tumour-suppressor gene called *APC*. If the developing polyp then suffers a second mutation jamming on an oncogene called *RAS*, it develops into a so-called 'adenoma'. If it then suffers a third mutation breaking another, unidentified tumour-suppressor gene, the adenoma grows into a more serious tumour. And now comes the danger of a fourth mutation, in the *TP53* gene, which turns the tumour into a full carcinoma. Similar multi-hit models apply to other kinds of cancer, with *TP53* often coming last.

You can now see why detecting cancer early in the development of the tumour is so important. The larger a tumour becomes, the more likely it is to suffer the next mutation, both because of general probability and because the rapid proliferation of cells inside the tumour can easily lead to genetic mistakes, which can cause mutations. People who are especially susceptible to certain cancers often carry mutations in 'mutator' genes, which encourage mutation generally (the breast cancer genes *BRCA1* and *BRCA2*, discussed in the chapter on chromosome 13, are probably breast-specific mutator genes), or because they already carry one faulty version of a tumour-suppressor gene. Tumours, like populations of rabbits, are prone to rapid and strong evolutionary pressures. Just as the offspring of the fastest-breeding rabbits soon dominate a rabbit warren, so the fastest dividing cells in each tumour come to dominate at the expense of more stable cells. Just as mutant rabbits that burrow underground to escape buzzards soon come to dominate at the expense of rabbits that sit in the open, so mutations in tumour-suppressor genes that enable cells to escape suppression soon come to dominate at the expense of other mutations. The environment of the tumour is literally selecting for mutations in such genes as the external environment selects rabbits. It is not mysterious that mutations eventually show up in so many cases. Mutation is random, but selection is not.

Likewise, it is now clear why cancer is a disease that very roughly doubles in frequency every decade of our lives, being principally a disease of old age. In somewhere between a tenth and a half of us, depending on the country we live in, cancer will eventually get round the various tumour-suppressor genes, including *TP53*, and will inflict a terrible and possibly fatal disease upon us. That this is a sign of the success of preventative medicine, which has eliminated so many other causes of death at least in the industrialised world, is little consolation. The longer we live, the more mistakes we accumulate in our genes, and the greater the chance that an oncogene may be jammed on and three tumour-suppressor genes jammed off in the same cell. The chances of this occurring are almost unimaginably small, but then the number of cells we make in our lifetimes is almost unimaginably large. As Robert Weinberg has put it:[5] 'One fatal malignancy per one hundred million billion cell divisions does not seem so bad after all.'

Let us take a closer look at the *TP53* gene. It is 1,179 'letters' long, and encodes the recipe for a simple protein, p53, that is normally rapidly digested by other enzymes so that it has a half-life of only twenty minutes. In this state, p53 is inactive. But upon receipt of a signal, production of the protein increases rapidly and destruction of it almost ceases. Exactly what that signal is remains shrouded in mystery and confusion, but damage to DNA is part of it. Bits of broken DNA seem somehow to alert p53. Like a criminal task force or SWAT team, the molecule scrambles to action stations. What happens next is that p53 takes charge of the whole cell, like one of those characters played by Tommy Lee Jones or Harvey Keitel who arrives at the scene of an incident and says something like: 'FBI: we'll take over from here.' Mainly by switching on other genes, p53 tells the cell to do one of two things: either to halt proliferation, stop replicating its DNA and pause until repaired; or to kill itself.

Another sign of trouble that alerts p53 is if the cell starts to run short of oxygen, which is a diagnostic feature of tumour cells. Inside a growing ball of cancer cells, the blood supply can run short, so the cells begin to suffocate. Malignant cancers get over this problem by sending out a signal to the body to grow new arteries into the tumour – the characteristic, crab-claw-like arteries that first gave cancer its Greek name. Some of the most promising new cancer drugs block this process of 'angiogenesis', or blood-vessel formation. But p53 sometimes realises what is happening and kills the tumour cells before the blood supply arrives. Cancers in tissues with poor blood supply, such as skin cancers, therefore must disable *TP53* early in their development or fail to grow. That is why melanomas are so dangerous.[6]

Little wonder that p53 has earned the nickname 'Guardian of the Genome', or even 'Guardian Angel of the Genome'. *TP53* seems to encode the greater good, like a suicide pill in the mouth of a soldier that dissolves only when it detects evidence that he is about to mutiny. The suicide of cells in this way is known as apoptosis, from the Greek for the fall of autumn leaves. It is the most important of the body's weapons against cancer, the last line of defence. Indeed, so important is apoptosis that it is gradually becoming clear that almost all therapeutic cancer treatment works only because it induces apoptosis by alerting p53 and its colleagues. It used to be thought that radiation therapy and chemotherapy worked because they preferentially killed dividing cells by damaging their DNA as it was being copied. But if that is the case, why do some tumours respond so poorly to treatment? There comes a point in the progression of fatal cancer when the treatment no longer works – the tumour no longer shrinks under chemical or radiation attack. Why should this be? If the treatment kills dividing cells, it should continue to work at all times.

Scott Lowe, working at Cold Spring Harbor Laboratory, has an ingenious answer. These treatments do indeed cause a little DNA damage, he says, but not enough to kill the cells. Instead, the DNA damage is just sufficient to alert p53, which tells the cells to commit suicide. So chemotherapy and radiation therapy are actually, like vaccination, treatments that work by helping the body to help itself. The evidence for Lowe's theory is good. Radiation, or treatment with 5-fluorouracil, etoposide or adriamycin – three chemical cancer treatments – all encourage apoptosis in laboratory cells infected with a viral oncogene. And when hitherto tractable tumours relapse and suddenly fail to respond to treatment, the change correlates closely with a mutation knocking out *TP53*. Likewise, the most intractable tumours – melanoma, lung, colorectal, bladder and prostate – are the ones in which *TP53* is usually mutated already. Certain kinds of breast cancer resist treatment: the ones in which *TP53* is broken.

These insights are of great importance to the treatment of cancer. A major branch of medicine has been acting under a large misapprehension. Instead of looking for agents that kill dividing cells, doctors should have been looking for agents that encourage cell suicide. That does not mean chemotherapy has been wholly ineffective, but it has been effective only by accident. Now that medical research knows what it is doing, the results should be more promising. In the short term it promises a less painful death for many cancer patients. By testing to see if *TP53* is already broken, doctors should soon be able to tell in advance if chemotherapy will work. If it will not, then the patient and his or her family can be spared the suffering and false hope that is now such a feature of the last months of life for such people.[7]

Oncogenes, in the unmutated state, are needed for cells to grow and proliferate normally throughout life: skin must be replaced, new blood cells generated, wounds repaired and so on. The mechanism for suppressing potential cancers must allow exceptions for normal growth and proliferation. Cells must frequently be given permission to divide, and must be equipped with genes that encourage division, so long as they stop at the right moment. How this feat is achieved is beginning to become clear. If we were looking at a man-made thing, we would conclude that a fiendishly ingenious mind must be behind it.

Once again, the key is apoptosis. Oncogenes are genes that cause division and growth but, surprisingly, several of them also trigger cell death. In the case of one such gene, known as *MYC*, both division and death are triggered by the gene, but its death signal is temporarily suppressed by external factors called survival signals. When these survival signals run out, death takes over. It is as if the designer, aware of *MYC*'s capacity to run amok, has automatically booby-trapped it so that any cell which gets carried away kills itself as soon as the supply of survival factors expires. The ingenious designer has gone a step further, by tying together three different oncogenes, *MYC*, *BCL-2* and *RAS*, so that they control each other. Normal cell growth can only occur if all three are working properly. In the words of the scientists who discovered these connections:[8] 'Without such support, booby traps spring and the affected cell is either killed or rendered moribund – either way, it is no longer a [cancerous] threat.'

The story of p53 and the oncogenes, like much of my book, challenges the argument that genetic research is necessarily dangerous and should be curtailed. The story also strongly challenges the view that 'reductionist' science, which takes systems apart to understand them, is flawed and futile. Oncology, the medical study of whole cancers, diligent, brilliant and massively endowed though it was, achieved terribly little by comparison with what has already been achieved in a few years by a reductionist, genetic approach. Indeed, one of the first calls for the complete sequencing of the human genome came from the Italian Nobel prize-winner Renato Dulbecco in 1986 because, he argued, it was the only way to win the war on cancer. There is now, for the first time in human history, a real prospect of a genuine cure for cancer, the cruellest and most common killer of all in the west, and it has come from reductionist, genetic research and the understanding that this brings. Those who would damn the whole science as dangerous should remember that.[9]

Natural selection, once she has selected a method of solving one problem, frequently uses it to solve another. Apoptosis has other functions than the elimination of cancer cells. It is also useful in the fight against ordinary infectious disease. If a cell detects that it has been infected with a virus, it can kill itself for the good of the body as a whole (ants and bees may do this as well, for the good of their colonies). There is good evidence that some cells do indeed do exactly this. There is also, inevitably, evidence that some viruses have evolved a way of preventing this from happening. Epstein–Barr virus, the cause of glandular fever or mononucleosis, contains a latent membrane protein whose job seems to be to head off any tendency the infected cell shows to commit suicide. Human papilloma virus, cause of cervical cancer, has two genes aboard whose job is to switch off *TP53* and another tumour-suppressor gene.

As I mentioned in the chapter on chromosome 4, Huntington's disease consists of unplanned and excessive apoptosis of brain cells which cannot then be replaced. Neurons cannot be regenerated in the adult brain – which is why some brain damage is irreversible. This makes good evolutionary sense because unlike, say, skin cells, each neuron is an exquisitely shaped, trained and experienced operator. To replace it with a naïve and untrained randomly shaped neuron would be worse than useless. When a virus gets into a neuron, the neuron is not instructed to undergo apoptosis. Instead, for reasons that are not entirely clear, the virus itself sometimes induces apoptosis of the neuron. This is true in the case of fatal alphavirus encephalitis, for instance.[10]

Apoptosis can also be useful in preventing other kinds of mutiny than cancer, such as genetic distortion of the kind induced by selfish transposons. There is some good evidence that the germ cells in the ovary and testicle are under surveillance from follicular and Sertoli cells respectively, whose job is to detect any such selfishness and, if so, to induce apoptosis. In the ovary of a five-month-old human foetus, for example, there are nearly seven million germ cells. By birth, there are only two million, and of those two million, just 400 or so will be ovulated during the coming lifetime. Most of the rest will be culled by apoptosis, which is ruthlessly eugenic, issuing strict orders to cells that are not perfect to commit suicide (the body is a totalitarian place).

The same principles may apply in the brain, where there is mass culling of cells during development by *ced-9* and other genes. Again, any cell that does not work well is sacrificed for the good of the whole. So not only does the apoptotic cull of neurons enable learning to take place, it also improves the average quality of the cells that remain. Something similar probably happens in the immune cells, another subject to ruthless culling of cells by apoptosis.

Apoptosis is a decentralised business. There is no central planning, no bodily Politburo deciding who should die and who should live. That is the beauty of it. Like the development of the embryo, it harnesses the self-knowledge of each cell. There is only one conceptual difficulty: how apoptosis could have evolved. In passing the test of killing itself if infected, cancerous or genetically mischievous, a cell by definition dies. It cannot therefore pass on its goodness to its daughters. Known as 'the kamikaze conundrum', this problem is solved by a form of group selection: whole bodies in which apoptosis works well do better than whole bodies in which it fails to work; the former therefore pass on the right traits to the cells of their offspring. But it does mean that the apoptotic system cannot improve during a person's lifetime, because it cannot evolve by natural selection within the body. We are stuck with the cell-suicide mechanism that we inherited.[11]

CHROMOSOME 18

Cures

Our doubts are traitors,
And make us lose the good we oft might win,
By fearing to attempt.
William Shakespeare, Measure for Measure

As the third millennium dawns, we are for the first time in a position to edit the text of our genetic code. It is no longer a precious manuscript; it is on disc. We can cut bits out, add bits in, rearrange paragraphs or write over words. This chapter is about how we can do these things, whether we should, and why, on the brink of doing so, our courage seems to be failing us and we are strongly tempted to throw away the whole word processor and insist that the text remains sacrosanct. This chapter is about genetic manipulation.

For most laymen, the obvious destination towards which genetic research is headed, the ultimate prize if you like, is a genetically engineered human being. One day, centuries hence, that might mean a human being with newly invented genes. For the moment it means a human being with an existing gene borrowed from another human being, or from an animal or plant. Is such a thing possible? And, if it is possible, is it ethical?

Consider a gene on chromosome 18 that suppresses colon cancer. We have already met it briefly in the last chapter: a tumour suppressor whose location has not quite been determined for sure. It was thought to be a gene called *DCC*, but we now know that *DCC* guides the growth of nerves in the spinal column and has nothing to do with tumour suppression. The tumour-suppressor gene is close to *DCC*, but it is still elusive. If you are born with an already faulty version of this gene you have a much increased risk of cancer. Could a future genetic engineer take it out, like a faulty spark plug from a car, and replace it? The answer, quite soon, will be yes.

I am just old enough to have begun my career in journalism cutting paper with real scissors and pasting with real glue. Nowadays, to move paragraphs around I use little software icons suitably decorated by the kind folk at Microsoft to indicate that they do the same job. (I have just moved this paragraph to here from the next page.) But the principle is the same: to move text, I cut it out and paste it back in somewhere else.

To do the same for the text of genes also requires scissors and glue. In both cases, fortunately nature had already invented them for her own purposes. The glue is an enzyme called ligase, which stitches together loose sentences of DNA whenever it comes across them. The scissors, called restriction enzymes, were discovered in bacteria in 1968. Their role in the bacterial cell is to defeat viruses by chopping up their genes. But it soon emerged that, unlike real scissors, a restriction enzyme is fussy: it only cuts a strand of DNA where it encounters a particular sequence of letters. We now know of 400 different kinds of restriction enzymes, each of which recognises a different sequence of DNA letters and cuts there, like a pair of scissors that cuts the paper only where it finds the word 'restriction'.

In 1972, Paul Berg of Stanford University used restriction enzymes in a test tube to chop two bits of viral DNA in half, then used ligases to stick them together again in new combinations. He thus produced the first man-made 'recombinant' DNA. Humanity could now do what retroviruses had long been doing: insert a gene into a chromosome. Within a year, the first genetically engineered bacterium existed: a gut bacterium infected with a gene taken out of a toad.

There was an immediate surge of public concern and it was not confined to lay people. Scientists themselves thought it right to pause before rushing to exploit the new technology. They called a moratorium on all genetic engineering in 1974, which only fanned the flames of public worry: if the scientists were worried enough to stop, then there really must be something to worry about. Nature placed bacterial genes in bacteria and toad genes in toads; who were we to swap them? Might the consequences not be terrible? A conference, held at Asilomar in 1975, thrashed out the safety arguments and led to a cautious resumption of genetic engineering in America under the supervision of a federal committee. Science was policing itself. The public anxiety seemed gradually to die down, though it was to revive quite suddenly in the mid-1990s, this time focusing not on safety but on ethics.

Biotechnology was born. First Genentech, then Cetus and Biogen, then other companies sprang up to exploit the new technique. A world of possibilities lay before the nascent businesses. Bacteria could now be induced to make human proteins for medicine, food or industrial use. Only gradually did disappointment dawn, when it emerged that bacteria were not very good at making most human proteins and that human proteins were too little known to be in great demand as medicines. Despite immense venture-capital investment, the only companies that made profits for their shareholders were the ones, such as Applied Biosystems, that made equipment for the others to use. Still, there were products. By the late 1980s, human growth hormone, made by bacteria, had replaced the expensive and dangerous equivalent extracted from the brains of cadavers. The ethical and safety fears proved so far groundless: in thirty years of genetic engineering no environmental or public health accident big or small has resulted from a genetic engineering experiment. So far, so good.

Meanwhile, genetic engineering had a greater impact on science than it had on business. It was now possible to 'clone' genes (in this context the word has a different meaning from the popular one): to isolate a 'needle' of a human gene in the 'haystack' that is the genome, put it in a bacterium and grow millions of copies of it so that they can be purified and the sequence of letters in the gene read. By this means, vast libraries of human DNA have been created containing thousands of overlapping fragments of the human genome, each present in sufficient quantity to study.

It is from such libraries that the people behind the Human Genome Project are piecing together the complete text. The scale of their task is immense. Three billion letters of text would fill a stack of books 150 feet high. The Wellcome Trust's Sanger Centre near Cambridge, which leads the effort, is reading the genome at the rate of 100 million letters a year.

There are, of course, short cuts. One is to ignore the ninety-seven per cent of the text that is silent – the selfish DNA, the introns, repetitive minisatellites and rusting pseudogenes – and concentrate on the genes alone. The quickest way to find such genes is to clone a different sort of library, called a cDNA library. First, sieve out all fragments of RNA in the cell. Many of them will be messengers – edited and abridged copies of genes in the process of being translated. Make DNA copies of those messengers and you will have, in theory, copies of the texts of the original genes with none of the junk DNA that lies in between. The main difficulty with this approach is that it gives no hint of the order or position of the genes on the chromosomes. By the late 1990s there was a marked difference of opinion between those who wanted to pursue this 'shotgun' method to the human genome with commercial patenting along the way, and those who wanted to be slow, thorough and public. On one side was a high-school drop-out, former professional surfer, Vietnam veteran and biotechnology millionaire named Craig Venter, backed by his own company Celera; on the other a studious, bearded, methodical Cambridge-educated scientist, John Sulston, backed by the medical charity Wellcome Trust. No prizes for guessing which camp is which.

But back to manipulation. Engineering a gene into a bacterium is one thing; inserting a gene into a human being is quite another. Bacteria are happy to absorb little rings of DNA called plasmids and adopt them as their own. Moreover, each bacterium is a single cell. Human beings have 100 trillion cells. If your goal is to genetically manipulate a human being, you need to insert a gene into every relevant cell, or start with a single-celled embryo.

The discovery in 1970 that retroviruses could make DNA copies from RNA suddenly made 'gene therapy' seem, nonetheless, a feasible goal. A retrovirus contains a message written in RNA which reads, in essence: 'Make a copy of me and stitch it into your chromosome.' All a gene therapist need do is take a retrovirus, cut out a few of its genes (especially those that make it infectious after the first insertion), put in a human gene, and infect the patient with it. The virus goes to work inserting the gene into the cells of the body and, lo, you have a genetically modified person.

Throughout the early 1980s, scientists worried about the safety of such a prodecure. The retrovirus might work too well and infect not just the ordinary cells of the body, but the reproductive cells, too. The retrovirus might reacquire its missing genes somehow and turn virulent; or it might destabilise the body's own genes and trigger cancer. Anything might happen. Fears about gene therapy were inflamed in 1980 when Martin Cline, a scientist studying blood disorders, broke a promise not to try inserting a harmless recombinant gene into an Israeli suffering from the genetic blood disorder thalassaemia (though not by retrovirus). Cline lost his job and his reputation; the result of his experiment was never published. Everybody agreed that human experiments were premature, to say the least.

But mouse experiments were proving both reassuring and disappointing. Far from being unsafe, gene therapy seemed more likely to be unworkable. Each retrovirus can only infect one kind of tissue; it needs careful packaging to get the genes into its envelope; it lands at random anywhere among the chromosomes and often fails to get switched on; and the body's immune system, primed by the crack troops of infectious disease, does not miss a clumsy, home-made retrovirus. Moreover, by the early 1980s so few human genes had been cloned that there was no obvious candidate gene to put in a retrovirus even if it could be got to work.

None the less, by 1989 several milestones had been passed. Retroviruses had carried rabbit genes into monkey cells; they had put cloned human genes into human cells; and they had put cloned human genes into mice. Three bold, ambitious men – French Anderson, Michael Blaese and Steven Rosenberg – decided the time was ripe for a human experiment. In a long and sometimes bitter battle with the American federal government's Recombinant DNA Advisory Committee, they sought permission for an experiment on terminal cancer patients. The argument brought out the different priorities of scientists and doctors. To the pure scientists, the experiment seemed hasty and premature. To the doctors, used to watching patients die of cancer, haste comes naturally. 'What's the rush?' asked Anderson at one session. 'A patient dies of cancer every minute in this country. Since we began this discussion 146 minutes ago, 146 patients have died of cancer.' Eventually, on 20 May 1989, the committee granted permission and two days later Maurice Kuntz, a truck driver dying from melanoma, received the first deliberately introduced (and approved) new gene. It was not designed to cure him, nor even to remain in his body permanently. It was simply an adjunct to a new form of cancer treatment. A special kind of white blood cell, good at infiltrating tumours and eating them, had been cultivated outside his body. Before injecting them back in, the doctors infected them with retroviruses carrying a little bacterial gene, the only purpose of which was to enable them to track the cells inside his body and find out where they went. Kuntz died, and nothing very surprising emerged from the experiment. But gene therapy had begun.

By 1990, Anderson and Blaese were back before the committee with a more ambitious scheme. This time the gene would actually be a cure, rather than just an identification tag. The target was an extremely rare inherited disease called severe combined immune deficiency (SCID), which rendered children incapable of mounting an immune defence against infection; the cause was the rapid death of all white blood cells. Unless kept in a sterile bubble or given a complete bone marrow transplant from a fortuitously matched relative, such a child faces a short life of repeated infection and illness. The disease is caused by a 'spelling' change in a single gene on chromosome 20, called the *ADA* gene.

Anderson and Blaese proposed to take some white blood cells from the blood of a SCID child, infect them with a retrovirus armed with a new *ADA* gene, and transfuse them back into the child's body. Once again, the proposal ran into trouble, but this time the opposition came from a different direction. By 1990, there was a treatment for SCID, called PEG-ADA, and it consisted of ingeniously delivering into the blood not the *ADA* gene, but ADA itself, the protein made by the equivalent gene in cattle. Like the cure for diabetes (injected insulin) or for haemophilia (injected clotting agents), SCID had been all but cured by protein therapy (injected PEG-ADA). What need was there of gene therapy?

At their birth, new technologies often seem hopelessly uncompetitive. The first railways were far more expensive than the existing canals and far less reliable. Only gradually and with time does the new invention bring down its own costs or raise its efficacy to the point where it can match the old. So it was with gene therapy. Protein therapy had won the race to cure SCID, but it required painful monthly injections into the hip, it was expensive and it needed to continue for life. If gene therapy could be made to work, it would replace all that with a single treatment that re-equipped the body with the gene it should have had in the first place.

In September 1990, Anderson and Blaese got the go-ahead and treated Ashanthi DeSilva, a three-year-old girl, with genetically engineered *ADA*. It was an immediate success. Her white-cell count trebled, her immunoglobulin counts soared and she began making almost a quarter of the ADA that an average person makes. The gene therapy could not be said to have cured her, because she was already receiving and continued to receive PEG-ADA. But gene therapy had worked. Today more than one in four of all known

SCID children in the world have had gene therapy. None are definitively cured enough to be weaned off PEG-ADA, but the side-effects have been few.

Other conditions will soon join SCID on the list of disorders that have been tackled by retroviral gene therapy, including familial hypercholesterolaemia, haemophilia and cystic fibrosis. But it is cancer that is undoubtedly the main target. In 1992 Kenneth Culver tried an audacious experiment that involved the first direct injection of gene-equipped retroviruses into the human body (as opposed to infection of cultured cells outside the body and transfusion of those cells back in). He injected retroviruses directly into brain tumours of twenty people. Injecting anything into the brain sounds horrifying enough, let alone a retrovirus. But wait till you hear what was in the retrovirus. Each one was equipped with a gene taken from a herpes virus. The tumour cells took up the retrovirus and expressed the herpes gene. But by then the cunning Dr Culver was treating the patient with drugs for herpes; the drugs attacked the tumours. It seemed to work on the first patient, but on four of the next five it failed.

These are early days in gene therapy. Some think it will one day be as routine as heart transplants are today. But it is too early to tell if gene therapy will be the strategy that defeats cancer, or whether some treatment based on blocking angiogenesis, telomerase or p53 wins that particular race. Whichever, never in history has cancer treatment looked so hopeful – thanks almost entirely to the new genetics.[1]

Somatic gene therapy of this kind is no longer very controversial. Concerns about safety still remain, of course, but almost nobody can think of an ethical objection. It is just another form of therapy and nobody who has watched a friend or relative go through chemotherapy or radiotherapy for cancer would begrudge them, on far-fetched safety grounds, the comparatively painless possibility of gene therapy instead. The added genes go nowhere near the germ cells that will form the next generation; that worry has been firmly laid to rest. Yet germline gene therapy – changing genes in places where they would be passed on to future generations, which remains a total taboo in human beings – would in one sense be much, much easier to do. It is germline gene therapy, in the form of genetically modified soya beans and mice, that has caused a resurgence of protest in the 1990s. This is, to borrow a term from its detractors, Frankenstein technology.

The genetic engineering of plants took off rapidly for several reasons. The first was commercial: farmers have for many years provided an eager market for new seed varieties. In ancient pre-history, conventional breeding had turned wheat, rice and maize from wild grasses to productive crops entirely by manipulating their genes, though these early farmers did not of course know that this is what they were doing. In modern times, the same techniques have trebled yields and increased per-capita food production by more than twenty per cent even as world population doubled between 1960 and 1990. The 'green revolution' in tropical agriculture was largely a genetic phenomenon. Yet all this had been done blindly: how much more could be achieved by targeted, careful gene manipulation? The second reason for the genetic engineering of plants is the ease with which plants can be cloned or propagated. You cannot take a cutting from a mouse and grow a new mouse as you can from many plants. But the third reason was a lucky accident. A bacterium called *Agrobacterium* had already been discovered, which had the unusual property of infecting plants with small loops of DNA called Ti plasmids that incorporated themselves into plant chromosomes. *Agrobacterium* was a ready-made vector: simply add some genes to the plasmid, rub it on a leaf, wait for the infection to take hold and grow a new plant from the leaf cells. The plant would now pass on the new gene in its seeds. So in 1983, first a tobacco plant, then a petunia and then a cotton plant were genetically modified in this way.

The cereals, which are resistant to *Agrobacterium* infection, had to wait until the invention of a rather more crude method: the genes are literally shot into the cell on board tiny particles of gold using gunpowder or particle accelerators. This technique has now become standard for all plant genetic engineering. It has led to the creation of tomatoes less likely to rot on the shelf, cotton resistant to boll weevils, potatoes resistant to Colorado beetles, maize resistant to corn borers and many other genetically modified plants.

The plants progressed from laboratory to field trial to commercial sale with relatively few hiccoughs. Sometimes the experiments did not work – boll weevils devastated the supposedly resistant cotton crop in 1996 – and sometimes they attracted protest from environmentalists. But there was never an 'accident'. When the genetically modified crops were brought across the Atlantic, they encountered stronger environmental resistance. In Britain in particular, where food safety regulators had lost public confidence after the 'mad-cow' epidemic, genetically modified food was suddenly a big issue in 1999, three years after it had become routine in the United States. Moreover, in Europe Monsanto made the mistake of starting with crops rendered resistant to its own indiscriminate herbicide, Roundup. This enabled the farmer to use Roundup to kill weeds. Such a combination of manipulating nature, encouraging use of herbicides and making profits infuriated many environmentalists. Eco-terrorists began tearing up experimental plots of genetically manipulated oilseed rape and paraded around in Frankenstein suits. The issue became one of Greenpeace's top three concerns, a sure sign of populism.

The media, as usual, rapidly polarised the debate with shouting matches between extremists on late-night television and interviews that forced people into simplistic answers: are you for or against genetic engineering? The issue reached its nadir when a scientist was forced to take early retirement over claims made in a hysterical television programme that he had proved that potatoes into which lectin genes had been inserted were bad for rats; he was later 'vindicated' by a group of colleagues assembled by Friends of the Earth. The result proved less about the safety of genetic engineering than it did about the safety of lectins – known animal poisons. The medium had become confused with the message. Putting arsenic in a cauldron makes the stew poisonous, but it does not mean all cooking is dangerous.

In the same way, genetic engineering is as safe and as dangerous as the genes that are engineered. Some are safe, some are dangerous. Some are green, some are bad for the environment. Roundup-resistant rape may be eco-unfriendly to the extent that it encourages herbicide use or spreads its resistance to weeds. Insect-resistant potatoes are eco-friendly to the extent that they require fewer insecticide applications, less diesel for the tractors applying the insecticides, less road use by the trucks delivering the insecticides and so on. The opposition to genetically modified crops, motivated more by hatred of new technology than love of the environment, largely chooses to ignore the fact that tens of thousands of safety trials have been done with no nasty surprises; that gene swapping between different species, especially microbes, is now known to be far more common than was once believed, so there is nothing 'unnatural' about the principle; that before genetic modification, plant breeding consisted of deliberate and random irradiation of seeds with gamma rays to induce mutations; that the main effect of genetic modification will be to reduce dependence on chemical sprays by improving resistance to diseases and pests; and that fast increases in yields are good for the environment, because they take the pressure off the cultivation of wild land.

The politicisation of the issue has had absurd results. In 1992, Pioneer, the world's biggest seed company, introduced a gene from brazil nuts into soya beans. The purpose was to make soya beans more healthy for those for whom they are a staple food by correcting soya beans' natural deficiency in a chemical called methionine. However, it soon emerged that a very few people in the world develop an allergy to brazil nuts, so Pioneer tested its transgenic soya beans and they proved allergenic, too, to such people. At this point, Pioneer alerted the authorities, published the results and abandoned the project. This was despite the fact that calculations showed that the new soya-bean allergy would probably kill no more than two Americans a year and could save hundreds of thousands worldwide from malnutrition. Yet instead of becoming an example of extreme corporate caution, the story was repackaged by environmentalists and told as a tale of the dangers of genetic engineering and reckless corporate greed.[2]

None the less, and even allowing for the cautious cancellation of many projects, it is a safe estimate that by the year 2000, fifty to sixty per cent of the crop seed sold in the United States will be genetically modified. For better or for worse, genetically modified crops are here to stay.

So are genetically modified animals. Putting a gene into an animal so that it and its offspring are permanently altered is now simple in animals as well as plants. You just stick it in. Suck your gene into the mouth of a very fine glass pipette, jab the tip of the pipette into a single-celled mouse embryo, extracted from a mouse twelve hours after mating, make sure the tip of the pipette is inside one of the cell's two nuclei, and press gently. The technique is far from perfect: only about five per cent of the resulting mice will have the desired gene switched on, and in other animals such as cows, success is even rarer. But in those five per cent the result is a 'transgenic' mouse with the gene incorporated in a random position on one of its chromosomes.

Transgenic mice are scientific gold dust. They enable scientists to find out what genes are for and why. The inserted gene need not be derived from a mouse, but could be from a person: unlike in computers, virtually all biological bodies can run any kind of software. For instance, a mouse that is abnormally susceptible to cancer can be made normal again by the introduction of a human chromosome 18, which formed part of the early evidence for a tumour-suppressor gene on chromosome 18. But rather than inserting whole chromosomes, it is more usual to add a single gene.

Micro-injection is giving way to a subtler technique, which has one distinct advantage: it can enable the gene to be inserted in a precise location. At three days of age, the embryo of a mouse contains cells known as embryonic stem cells or ES cells. If one of these is extracted and injected with a gene, as Mario Capecchi was the first to discover in 1988, the cell will splice that gene in at precisely the point where the gene belongs, replacing the existing version of the gene. Capecchi took a cloned mouse oncogene called *int-*2, inserted it into a mouse cell by briefly opening the cell's pores in an electric field, and then observed as the new gene found the faulty gene and replaced it. This procedure, called 'homologous recombination', exploits the fact that the mechanism that repairs broken DNA often uses the spare gene on the counterpart chromosome as a template. It mistakes the new gene for the template and corrects its existing gene accordingly. Thus altered, an ES cell can then be placed back inside an embryo and grown into a 'chimeric' mouse – a mouse in which some of the cells contain the new gene.[3]

Homologous replication allows the genetic engineer not only to repair genes but to do the opposite: deliberately to break working genes, by inserting faulty versions in their place. The result is a so-called knockout mouse, reared with a single gene silenced, the better to reveal that gene's true purpose. The discovery of memory mechanisms (see the chapter on chromosome 16) owes much to knockout mice, as do other fields of modern biology.

Transgenic animals are useful not only to scientists. Transgenic sheep, cattle, pigs and chickens have commercial applications. Sheep have already been given the gene for a human clotting factor in the hope that it can be harvested from their milk and used to treat haemophiliacs. (Almost incidentally, the scientists who performed this procedure cloned the sheep Dolly and displayed her to an amazed world in early 1997.) A company in Quebec has taken the gene that enables spiders to make silk webs and inserted it into goats, hoping to extract raw silk protein from the goats' milk and spin it into silk. Another company is pinning its hope on hens' eggs, which it hopes to turn into factories for all sorts of valuable human products, from pharmaceuticals to food additives. But even if these semi-industrial applications fail, transgenic technology will transform animal breeding, as it is transforming plant breeding, generating beef cattle that put on more muscle, dairy cattle that give more milk or chickens that lay tastier eggs.[4]

It all sounds rather easy. The technical obstacles to breeding a transgenic or a knockout human being are becoming trivial for a good team at a well-equipped laboratory. In a few years from now you probably could, in principle, take a complete cell from your own body, insert a gene into a particular location on a particular chromosome, transfer the nucleus to an egg cell from which the nucleus had been removed, mix the ensuing cell with a human embryo cloned from your body and grow a new chimeric human being from the embryo. The person would be a transgenic clone of yourself, identical in every way except, say, in having an altered version of the gene that made you go bald. You could alternatively use ES cells from such a clone to grow a spare liver to replace the one you sacrificed to the bottle. Or you could grow human neurons in the laboratory to test new drugs on, thus sparing the lives of laboratory animals. Or, if you were barking mad, you could leave your property to your clone and commit suicide secure in the knowledge that something of you still existed, but slightly improved. Nobody need know that this person is your clone. If the increasing resemblance to you later became apparent as he grew older, the non-receding hairline would soon lay suspicions to rest.

None of this is yet possible – human ES cells have only just been found – but it is very unlikely to remain impossible for much longer. When human cloning is possible, will it be ethical? As a free individual, you own your own genome and no government can nationalise it, nor company purchase it, but does that give you the right to inflict it on another individual? (A clone is another individual.) Or to tamper with it? For the moment society seems keen to bind itself against such temptations, to place a moratorium on cloning or germline gene therapy and strict limits on embryonic research, to forego the medical possibilities in exchange for not risking the horrors of the unknown. We have drummed into our skulls with every science fiction film the Faustian sermon that to tamper with nature is to invite diabolic revenge. We have grown cautious. Or at least we have as voters. As consumers, we may well act differently. Cloning may well happen not because the majority approves, but because the minority acts. That, after all, was roughly what happened in the case of test-tube babies. Society never decided to allow them; it just got used to the idea that those who desperately wanted such babies were able to have them.

Meanwhile, in one of those ironies which modern biology supplies in abundance, if you have a faulty tumour-suppressor gene on chromosome 18, forget gene therapy. A much simpler preventive treatment may be at hand. New research suggests that for those with genes that increase their susceptibility to bowel cancer, a diet rich in aspirin and unripe bananas offers the promise of protection against the cancer. The diagnosis is genetic, but the cure is not. Genetic diagnosis followed by conventional cure is probably the genome's greatest boon to medicine.

Prevention

Ninety-nine per cent of people don't have an inkling
about how fast this revolution is coming.
Steve Fodor, president of Affymetrix

The improvement of any medical technology confronts our species
with a moral dilemma. If the technology can save lives, then not to
develop it and use it is morally culpable, even if there are attendant
risks. In the Stone Age, we had no option but to watch our relatives
die of smallpox. After Jenner had perfected vaccination we were
derelict in our duty if we did so. In the nineteenth century, we had
no alternative to watching our parents succumb to tuberculosis.
After Fleming found penicillin we were guilty of neglect if we failed
to take a dying tubercular patient to the doctor. And what applies
on the individual level applies with even greater force on the level
of countries and peoples. Rich countries can no longer ignore the
epidemics of diarrhoea that claim the lives of countless children in
poor countries, because no longer can we argue that nothing medi-
cally can be done. Oral rehydration therapy has given us a conscience.
Because something can be done, so something must be done.

This chapter is about the genetic diagnosis of two of the commonest diseases that afflict people, one a swift and merciless killer, the other a slow and relentless thief of memory: coronary heart disease and Alzheimer's disease. I believe we are in danger of being too squeamish and too cautious in using knowledge about the genes that influence both diseases, and we therefore stand at risk of committing the moral error of denying people access to life-saving research.

There is a family of genes called the apolipoprotein genes, or *APO* genes. They come in four basic varieties, called A, B, C and – strangely – E, though there are various different versions of each on different chromosomes. The one that interests us most is *APOE*, which happens to lie here on chromosome 19. To understand *APOE*'s job requires a digression into the habits of cholesterol and triglyceride fats. When you eat a plate of bacon and eggs, you absorb much fat and with it cholesterol, the fat-soluble molecule from which so many hormones are made (see the chapter on chromosome 10). The liver digests this stuff and feeds it into the bloodstream for delivery to other tissues. Being insoluble in water, both triglyceride fats and cholesterol have to be carried through the blood by proteins called lipoproteins. At the beginning of the journey, laden with both cholesterol and fats, the delivery truck is called VLDL, for very-low-density lipoprotein. As it drops off some of its triglycerides, it becomes low-density lipoprotein, or LDL ('bad cholesterol'). Finally, after delivering its cholesterol, it becomes high-density lipoprotein, HDL ('good cholesterol') and returns to the liver for a new consignment.

The job of *APOE*'s protein (called apo-epsilon) is to effect an introduction between VLDL and a receptor on a cell that needs some triglycerides; *APOB*'s job (or rather apo-beta's) is to do the same for the cholesterol drop-off. It is easy to see therefore that *APOE* and *APOB* are prime candidates for involvement in heart disease. If they are not working, the cholesterol and fat stay in the bloodstream and can build up on the walls of arteries as atherosclerosis. Knockout mice with no *APOE* genes get atherosclerosis even on a normal mouse diet. The genes for the lipoproteins themselves and for the receptors on cells can also affect the way in which cholesterol and fat behave in the blood and thereby facilitate heart attacks. An inherited predisposition to heart disease, called familial hypercholesterolaemia, results from a rare 'spelling change' in the gene for cholesterol receptors.[1]

What marks *APOE* out as special is that it is so 'polymorphic'. Instead of us all having one version of the gene, with rare exceptions, *APOE* is like eye colour: it comes in three common kinds, known as *E2*, *E3* and *E4*. Because these three vary in their efficiency at removing triglycerides from the blood, they also vary in their susceptibility to heart disease. In Europe, *E3* is both the 'best' and the commonest kind: more than eighty per cent of people have at least one copy of *E3* and thirty-nine per cent have two copies. But the seven per cent of people who have two copies of *E4* are at markedly high risk of early heart disease, and so, in a slightly different way, are the four per cent of people who have two copies of *E2*.[2]

But that is a Europe-wide average. Like many such polymorphisms, this one shows geographical trends. The further north in Europe you go, the commoner *E4* becomes, at the expense of *E3* (*E2* remains roughly constant). In Sweden and Finland the frequency of *E4* is nearly three times as high as in Italy. So, approximately, is the frequency of coronary heart disease.[3] Further afield, there are even greater variations. Roughly thirty per cent of Europeans have at least one copy of *E4*; Orientals have the lowest frequency at roughly fifteen per cent; American blacks, Africans and Polynesians, over forty per cent; and New Guineans, more than fifty per cent. This probably reflects in part the amount of fat and fatty meat in the diet during the last few millennia. It has been known for some while that New Guineans have little heart disease when they eat their traditional diet of sugar cane, taro and occasional meals of lean bush meat from possums and tree kangaroos. But as soon as they get jobs at strip mines and start eating western hamburgers and chips, their risk of early heart attacks shoots up – much more quickly than in most Europeans.[4]

Heart disease is a preventable and treatable condition. Those with the *E2* gene in particular are acutely sensitive to fatty and cholesterol-rich diets, or to put it another way, they are easily treated by being warned off such diets. This is extremely valuable genetic knowledge. How many lives could be saved, and early heart attacks averted, by simple genetic diagnosis to identify those at risk and target treatment at them?

Genetic screening does not automatically lead to such drastic solutions as abortion or gene therapy. Increasingly a bad genetic diagnosis can lead to less drastic remedies: to the margarine tub and the aerobics class. Instead of warning us all to steer clear of fatty foods, the medical profession must soon learn to seek out which of us could profit from such a warning and which of us can relax and hit the ice cream. This might go against the profession's puritanical instincts, but not against its Hippocratic oath.

However, I did not bring you to the *APOE* gene chiefly to write about heart disease, though I fear I am still breaking my rule by writing about another disease. The reason it is one of the most investigated genes of all is not because of its role in heart disease, but because of its pre-eminent role in a much more sinister and much less curable condition: Alzheimer's disease. The devastating loss of memory and of personality that accompanies old age in so many people – and that occurs in a few people when quite young – has been attributed to all sorts of factors, environmental, pathological and accidental. The diagnostic symptom of Alzheimer's is the appearance in brain cells of 'plaques' of insoluble protein whose growth damages the cell. A viral infection was once suspected to be the cause, as was a history of frequent blows to the head. The presence of aluminium in the plaques threw suspicion on aluminium cooking pots for a while. The conventional wisdom was that genetics had little or nothing to do with the disease. 'It is not inherited,' said one textbook firmly.

But as Paul Berg, co-inventor of genetic engineering, has said, 'all disease is genetic' even when it is also something else. Pedigrees in which Alzheimer's disease appeared with high frequency were eventually discovered among the American descendants of some Volga Germans and by the early 1990s at least three genes had been associated with early-onset Alzheimer's disease, one on chromosome 21 and two on chromosome 14. But a far more significant discovery in 1993 was that a gene on chromosome 19 seemed to be associated with the disease in old people and that Alzheimer's in the elderly might also have a partial genetic basis. Quite soon the culprit gene was discovered to be none other than *APOE* itself.[5]

The association of a blood-lipid gene with a brain disease should not have come as such a surprise as it did. After all, it had been noticed for some time that Alzheimer's victims quite often had high cholesterol. None the less, the scale of the effect came as a shock. Once again, the 'bad' version of the gene is *E4*. The chances of getting Alzheimer's are twenty per cent for those with no *E4* gene and the mean age of onset is eighty-four. For those with one *E4* gene, the probability rises to forty-seven per cent and the mean age of onset drops to seventy-five. For those with two *E4* genes, the probability is ninety-one per cent and the mean age of onset sixty-eight years. In other words, if you carry two *E4* genes (and seven per cent of Europeans do), just about the only thing that can prevent you getting Alzheimer's disease is premature death from some other cause. There will still be a few who escape either fate – indeed, one study found an eighty-six-year-old *E4/E4* man with all his wits – but they will be very few. In many people who show no symptoms of memory loss, the classic plaques of Alzheimer's are none the less present, and they are usually worse in *E4* carriers than *E3*. Those with at least one *E2* version of the gene are even less likely to get Alzheimer's than those with *E3* genes, though the difference is small. This is no accidental side-effect or statistical coincidence: this looks like something central to the mechanism of the disease.[6]

Recall that *E4* is rare among Oriental people, commoner among whites, commoner still among Africans and commonest in New Guinean Melanesians. It should follow that Alzheimer's obeys the same gradient, but it is not quite so simple. The relative risk of getting Alzheimer's is much higher for white $E4/E4$s than for black or Hispanic $E4/E4$s – compared with the risk for $E3/E3$s. Presumably, susceptibility to Alzheimer's is affected by other genes, which vary between different races. Also, $E4$'s effects seem to be more severe among women than men. Not only do more women than men get Alzheimer's, but females who are $E4/E3$ are just as much at risk as those who are $E4/E4$. Among men, having one $E3$ gene reduces risk.[7]

You may be wondering why $E4$ exists at all, let alone at such high frequencies. If it exacerbates both heart disease and Alzheimer's, it should surely have been driven extinct by the more benign $E3$ and $E2$ long ago. I'm tempted to answer the question by saying that high-fat diets were until recently so rare that the coronary side-effects were of little importance, while Alzheimer's disease is all but irrelevant to natural selection, since it not only happens to people who have long ago reared their own children to independence, but strikes at an age when most Stone-Age folk were long dead anyway. But I am not sure that is a good enough answer, because meaty and even cheesy diets have been around a long time in some parts of the world – long enough for natural selection to go to work. I suspect that $E4$ plays yet another role in the body, which we do not know about, and at which it is better than $E3$. Remember: GENES ARE NOT THERE TO CAUSE DISEASES.

The difference between $E4$ and the commoner $E3$ is that the 334th 'letter' in the gene is G instead of A. The difference between $E3$ and $E2$ is that the 472nd 'letter' is a G instead of an A. The effect is to give $E2$'s protein two extra cysteines and $E4$'s two extra arginines compared with each other, $E3$ being intermediate. These tiny changes in a gene that is 897 'letters' long are sufficient to alter the way $APOE$'s protein does its job. Quite what that job is remains obscure, but one theory is that it is to stabilise another protein called tau, which is supposed in turn to keep in shape the tubular 'skeleton' of a neuron. Tau has an addiction to phosphate, which prevents it doing its job; $APOE$'s job is to keep tau off the phosphate. Another theory is that $APOE$'s job in the brain is not unlike its job in the blood. It carries cholesterol between and within brain cells so they can build and repair their fat-insulated cell membranes. A third and more direct theory is that, whatever *APOE*'s job, the *E4* version has a special affinity for something called amyloid beta peptide, which is the substance that builds up inside neurons of Alzheimer's sufferers. Somehow, it assists the growth of these destructive plaques.

The details will matter one day, but for now the important fact is that we are suddenly in possession of a means of making predictions. We can test the genes of individuals and make very good forecasts about whether they will get Alzheimer's disease. The geneticist Eric Lander recently raised an alarming possibility. We now know that Ronald Reagan has Alzheimer's, and it seems likely in retrospect that he had the early stages of the disease when he was in the White House. Suppose that some enterprising but biased journalist, anxious to find some way of discrediting Reagan as a presidential candidate in 1979, had snatched a napkin on which Reagan had wiped his mouth and tested the DNA on it (gloss over the fact that the test was not then invented). Suppose he had discovered that this second-oldest-ever presidential candidate was very likely to develop the disease in his term of office and had printed this finding in his newspaper.

The story illustrates the dangers for civil liberties that genetic testing brings with it. When asked if we should offer *APOE* tests to individuals curious to know if they will get Alzheimer's, most in the medical profession say no. After cogitating on the issue recently, the Nuffield Council on Bioethics, Britain's leading think-tank on such matters, reached the same conclusion. To test somebody for a disease that is incurable is dubious at best. It can buy reassurance for those who find themselves with no *E4* gene, but at a terrible price: the almost-certain sentence to an incurable dementia for those with two *E4* genes. If the diagnosis were absolutely certain, then (as Nancy Wexler argued in the case of Huntington's – see the chapter on chromosome 4), the test could be even more devastating. On the other hand, it would at least not be misleading. But in cases where there is less certainty, such as the $APOE$ case, the test would be of still less value. You can still – if you are very lucky – have two E_4 genes and live to an old age with no symptoms, just as you can still – if you are very unlucky – have no E_4 genes and get Alzheimer's at sixty-five. Since a diagnosis of two E_4 genes is neither sufficient nor necessary to predict Alzheimer's, and since there is no cure, you should not be offered the test unless you are already symptomatic.

At first I found all these argument convincing, but now I am not so sure. After all, it has been considered ethical to offer people the test for the HIV virus if they want it, even though AIDS was (until recently) incurable. AIDS is not an inevitable outcome of HIV infection: some people survive indefinitely with HIV infection. True, there is in the case of AIDS the additional interest of society in preventing the spread of the infection, which does not apply to Alzheimer's disease, but it is the individual at risk we are considering here, not society at large. The Nuffield Council addresses this argument by implicitly making a distinction between genetic and other tests. To attribute a person's susceptibility to an illness to their genetic make-up distorts attitudes, argued the report's author, Dame Fiona Caldicott. It makes people believe wrongly that genetic influences are paramount and causes them to neglect social and other causes; that, in turn, increases the stigma attached to mental illness.[8]

This is a fair argument unfairly applied. The Nuffield Council is operating a double standard. 'Social' explanations of mental problems offered by psychoanalysts and psychiatrists are licensed to practise on the flimsiest of evidence, yet they are just as likely to stigmatise people as genetic ones. They continue to flourish while the great and the good of bioethics outlaw diagnoses that are supported by evidence merely because they are genetic explanations. In striving to find reasons to outlaw genetic explanations while allowing social ones to flourish, the Nuffield Council even resorted to calling the predictive power of the $APOE_4$ test 'very low' – bizarre wording for an eleven-fold difference in risk between the E_4/E_4s and the E_3/E_3s.[9] As John Maddox has commented,[10] citing *APOE* as a case in point, 'There are grounds for suspecting that physicians are not pursuing valuable opportunities out of diffidence at revealing unwelcome genetic information to their patients ... but diffidence can be taken too far.'

Besides, although Alzheimer's disease is incurable, there are already drugs that alleviate some of the symptoms and there may be precautions of uncertain value that people can take to head it off. Is it not better to know if one should take every precaution? If I had two E_4 genes, I might well want to know so that I could volunteer for trials of experimental drugs. For those who indulge in activities that raise their risk of Alzheimer's disease, the test certainly makes sense. It is, for example, now apparent that professional boxers who have two E_4 genes are at such risk of developing early Alzheimer's that boxers are indeed best advised to take a test and not box if they find themselves with two E4s. One in six boxers get Parkinson's disease or Alzheimer's – the microscopic symptoms are similar, though the genes involved are not – by the age of fifty, and many, including Mohammed Ali, suffer even younger. Among those boxers who do get Alzheimer's, the E_4 gene is unusually common, as it is among people who suffer head injury and later turn out to have plaques in their neurons.

What is true for boxers may be true for other sports in which the head is struck. Alerted by anecdotal evidence that many great footballers sink into premature senility in old age – Danny Blanchflower, Joe Mercer and Bill Paisley being sad, recent examples from British clubs – neurologists have begun to study the prevalence of Alzheimer's disease in such sportsmen. Somebody has calculated that a footballer on average heads the ball 800 times in a season; the wear and tear could be considerable. A Dutch study did indeed find worse memory loss in footballers than in other sportsmen and a Norwegian one found evidence of brain damage in footballers. Once more it is plausible that the E_4/E_4 homozygotes might benefit from at least knowing at the outset of their careers that they were specially at risk. As somebody who frequently hits his head on door frames because architects have not made them big enough for tall people to walk through, I wonder myself what my *APOE* genes looks like. Maybe I should have them tested.

Testing could be valuable in other ways. At least three new Alzheimer's drugs are in development and testing. One that is already here, tacrine, is now known to work better in those with *E3* and *E2* genes than in *E4* carriers. Again and again the genome drives home the lesson of our individuality. The diversity of humanity is its greatest message. Yet there is still a marked reluctance in the medical profession to treat the individual rather than the population. A treatment that is suitable for one person may not suit another. Dietary advice that could save one person's life might do no good at all to another. The day will come when a doctor will not prescribe you many kinds of medicine until he has checked which version of a gene or genes you have. The technology is already being developed, by a small Californian company called Affymetrix among others, to put a whole genome-full of genetic sequences on a single silicon chip. One day we might each carry with us exactly such a chip from which the doctor's computer can read any gene the better to tailor his prescription to us.[11]

Perhaps you have already sensed what the problem with this would be – and what is the real reason behind the experts' squeamishness about *APOE* tests. Suppose I do have *E4/E4* and I am a professional footballer. I therefore stand a much higher than average chance of contracting angina and premature Alzheimer's disease. Suppose that today, instead of going to see my doctor, I am going to see an insurance broker to arrange a new life-insurance policy to go with my mortgage, or to get health insurance to cover future illness. I am handed a form and asked to fill in questions about whether I smoke, how much I drink, whether I have AIDS and what I weigh. Do I have a family history of heart disease? – a genetic question. Each question is designed to narrow me down into a particular category of risk so that I can be quoted an appropriately profitable, but still competitive premium. It is only logical that the insurance company will soon ask to see my genes as well, to ask if

I am $E4/E4$, or if I have a pair of $E3$s instead. Not only does it fear that I might be loading up on life insurance precisely because I know from a recent genetic test that I am doomed, thus ripping it off as surely as a man who insures a building he plans to burn down. It also sees that it can attract profitable business by offering discounts to people whose tests prove reassuring. This is known as cherry picking, and it is exactly why a young, slim, heterosexual non-smoker already finds he can get life insurance cheaper than an old, plump, homosexual smoker. Having two $E4$ genes is not so very different.

Little wonder that in America health-insurance companies are already showing interest in genetic tests for Alzheimer's, a disease that can be very costly for them (in Britain, where health cover is basically free, the main concern is life insurance). But mindful of the fury the industry unleashed when it began charging homosexual men higher premiums than heterosexuals to reflect the risk of AIDS, the industry is treading warily. If genetic testing were to become routine for lots of genes, the entire concept of pooled risk, on which insurance is based, would be undermined. Once my exact fate is known, I would be quoted a premium that covered the exact cost of my life. For the genetically unfortunate, it might prove unaffordable: they would become an insurance underclass. Sensitive to these issues, in 1997 the insurance industry association in Britain agreed that for two years it would not demand genetic tests as a condition of insurance and would not (for mortgages smaller than £100,000) demand to know the results of genetic tests you may already have taken. Some companies went even further, saying that genetic tests were not part of their plans. But this shyness may not last.

Why do people feel so strongly about this issue, when it would in practice mean cheaper premiums for many? Indeed, unlike so many things in life, genetic good fortune is equitably distributed among the privileged as well as the less privileged – the rich cannot buy good genes and the rich spend more on insurance anyway. The answer, I think, goes to the heart of determinism. A person's decision to smoke and drink, even the decision that led to his catching AIDS, was in some sense a voluntary one. His decision to have two E_4 genes at the $APOE$ gene was not a decision at all; it was determined for him by nature. Discriminating on the basis of $APOE$ genes is like discriminating on the basis of skin colour or gender. A non-smoker might justifiably object to subsidising the premium of a smoker by being lumped with him in the same risk category, but if an E_3/E_3 objected to subsidising the premium of an E_4/E_4, he would be expressing bigotry and prejudice against somebody who was guilty of nothing but bad luck.[12]

The spectre of employers using genetic tests to screen potential staff is less fraught. Even when more tests are available, there will be few temptations for employers to use them. Indeed, once we get more used to the idea that genes lie behind susceptibilities to environmental risks, some tests might become good practice for employer and employee alike. In a job where there is some exposure to known carcinogens (such as bright sunlight – the job of lifeguard, say), the employer may in future be neglecting his duty of care to his workers if he employs people with faulty p53 genes. He might, on the other hand, be asking applicants to take a genetic test for more selfish motives: to select people with healthier dispositions or more outgoing personalities (exactly what job interviews are designed to do), but there are already laws against discrimination.

Meanwhile, there is a danger that the hobgoblin of genetic insurance tests and genetic employment tests will scare us away from using genetic tests in the interests of good medicine. There is, however, another hobgoblin that scares me more: the spectre of government telling me what I may do with my genes. I am keen not to share my genetic code with my insurer, I am keen that my doctor should know it and use it, but I am adamant to the point of fanaticism that it is my decision. My genome is my property and not the state's. It is not for the government to decide with whom I may share the contents of my genes. It is not for the government to decide whether I may have the test done. It is for me. There is a terrible, paternalist tendency to think that 'we' must have one policy on this matter, and that government must lay down rules about how much of your own genetic code you may see and whom you may show it to. It is yours, not the government's, and you should always remember that.

Politics

Oh! The roast beef of England,
And Old England's roast beef.
Henry Fielding,
The Grub Street Opera

The fuel on which science runs is ignorance. Science is like a hungry furnace that must be fed logs from the forests of ignorance that surround us. In the process, the clearing we call knowledge expands, but the more it expands, the longer its perimeter and the more ignorance comes into view. Before the discovery of the genome, we did not know there was a document at the heart of every cell three billion letters long of whose content we knew nothing. Now, having read parts of that book, we are aware of myriad new mysteries.

The theme of this chapter is mystery. A true scientist is bored by knowledge; it is the assault on ignorance that motivates him — the mysteries that previous discoveries have revealed. The forest is more interesting than the clearing. On chromosome 20 there lies as irritating and fascinating a copse of mystery as any. It has already yielded two Nobel prizes, merely for the revelation that it is there, but it stubbornly resists being felled to become knowledge. And, as if to remind us that esoteric knowledge has a habit of changing the world, it became one of the most incendiary political issues in science one day in 1996. It concerns a little gene called *PRP*.

The story starts with sheep. In eighteenth-century Britain, agriculture was revolutionised by a group of pioneering entrepreneurs, among them Robert Bakewell of Leicestershire. It was Bakewell's discovery that sheep and cattle could be rapidly improved by selectively breeding the best specimens with their own offspring to concentrate desirable features. Applied to sheep this inbreeding produced fast-growing, fat lambs with long wool. But it had an unexpected side-effect. Sheep of the Suffolk breed, in particular, began to exhibit symptoms of lunacy in later life. They scratched, stumbled, trotted with a peculiar gait, became anxious and seemed antisocial. They soon died. This incurable disease, called scrapie, became a large problem, often killing one ewe in ten. The scrapie followed Suffolk sheep, and to a lesser extent other breeds, to other parts of the world. Its cause remained mysterious. The disease did not seem to be inherited, but it could not be traced to another origin. In the 1930s, a veterinary scientist, testing a new vaccine for a different disease, caused a massive epidemic of scrapie in Britain. The vaccine had been made partly from the brains of other sheep and although it had been thoroughly sterilised in formalin, it retained some infectious strength. From then on it became the orthodox, not to say blinkered, view of veterinary scientists that scrapie, being transmissible, must be caused by a microbe.

But what microbe? Formalin did not kill it. Nor did detergents, boiling or exposure to ultraviolet light. The agent passed through filters fine enough to catch the tiniest viruses. It raised no immune response in infected animals and there was sometimes a long delay between injection of the agent and disease – though the delay was much shorter if the agent was injected directly into the brain. Scrapie threw up a baffling wall of ignorance that defeated a generation of determined scientists. Even when similar symptoms appeared in American mink farms and in wild elk and mule deer inhabiting particular national parks in the Rocky Mountains, the mystery only deepened. Mink proved resistant to sheep scrapie when experimentally injected. By 1962, one scientist had returned to the genetic hypothesis. Perhaps, he suggested, scrapie is an inherited but also transmissible disease, a hitherto unknown combination. There are plenty of inherited diseases, and contagious diseases in which inheritance determines susceptibility – cholera being a now classic case – but the notion that an infectious particle could somehow travel through the germline seemed to break all the rules of biology. The scientist, James Parry, was firmly put in his place.

About this time an American scientist, Bill Hadlow saw pictures of the damaged brains of scrapie-riddled sheep in an exhibit in the Wellcome Museum of Medicine in London. He was struck by their similarity to pictures he had seen from a very different place. Scrapie was about to get a lot more relevant to people. The place was Papua New Guinea, where a terrible debilitating disease of the brain, known as kuru, had been striking down large numbers of people, especially women, in one tribe known as the Fore. First, their legs began to wobble, then their whole bodies started to shake, their speech became slurred and they burst into unexpected laughter. Within a year, as the brain progressively dissolved from within, the victim would be dead. By the late 1950s, kuru was the leading cause of death among Fore women, and it had killed so many that men outnumbered women by three to one. Children also caught the disease, but comparatively few adult men.

This proved a crucial clue. In 1957 Vincent Zigas and Carleton Gajdusek, two western doctors working in the area, soon realised what had been happening. When somebody died, the body was ceremonially dismembered by the women of the tribe as part of the funeral ritual and, according to anecdote, eaten. Funereal cannibalism was well on the way to being stamped out by the government, and it had acquired sufficient stigma that few people were prepared to talk openly about it. This has led some to question whether it ever happened. But Gajdusek and others gathered sufficient eye-witness accounts to leave little doubt that the Fore were not lying when they described pre-1960 funeral rituals in Pidgin as 'katim na kukim na kaikai' – or cut up, cook and eat. Generally, the women and children ate the organs and brains; the men ate the muscle. This immediately suggested an explanation for kuru's pattern of appearance. It was commonest among women and children; it appeared among relatives of victims – but among married relations as well as blood relatives; and after cannibalism became illegal, the age of its victims steadily increased. In particular, Robert Klitzman, a student of Gajdusek's, identified three clusters of deaths, each of which included only those who attended certain funerals of kuru victims in the 1940s and 1950s. For instance, at the funeral of a woman called Neno in 1954, twelve of fifteen relatives who attended later died of kuru. The three who did not comprised somebody who died young of another cause, somebody who was forbidden by tradition to take part in the eating because she was married to the same man as the dead woman, and somebody who later claimed to have eaten only a hand.

When Bill Hadlow saw the similarity between kuru-riddled human brains and scrapie-riddled sheep brains, he immediately wrote to Gajdusek in New Guinea. Gajdusek followed up the hint. If kuru was a form of scrapie, then it should be possible to transmit it from people to animals by direct injection into the brain. In 1962 his colleague, Joe Gibbs, began a long series of experiments to try to infect chimpanzees and monkeys with kuru from the brains of dead Fore people (whether such an experiment would now be regarded as ethical is outside the scope of this book). The first two chimpanzees sickened and died within two years of the injections. Their symptoms were like those of kuru victims.

Proving that kuru was a natural human form of scrapie did not help much, since scrapie studies were in such confusion over what could be the cause. Ever since 1900, a rare and fatal human brain disease had been teasing neurologists. The first case of what came to be known as Creutzfeldt–Jakob disease, or CJD, was diagnosed by Hans Creutzfeldt in Breslau in that year in an eleven-year-old girl who died slowly over the succeeding decade. Since CJD almost never strikes the very young and rarely takes so long to kill, this was almost certainly a strange case of misdiagnosis at the outset, leaving us with a paradox all too typical of this mysterious disease: the first CJD patient ever recognised did not have CJD. However, in the 1920s, Alfons Jakob did find cases of what probably was CJD and the name stuck.

Gibbs's chimpanzees and monkeys soon proved just as susceptible to CJD as they had been to kuru. In 1977, events took a more frightening turn. Two epileptics who had undergone exploratory brain surgery with electrodes at the same hospital suddenly developed CJD. The electrodes had been previously used in a CJD patient, but they had been properly sterilised after use. Not only did the mysterious entity that caused the disease resist formalin, detergent, boiling and irradiation, it survived surgical sterilisation. The electrodes were flown to Bethesda to be used on chimps, who promptly got CJD, too. This proved the beginning of a new and yet more bizarre epidemic: iatrogenic ('doctor-caused') CJD. It has since killed nearly one hundred people who had been treated for small stature with human growth hormone prepared from the pituitary glands of cadavers. Because several thousand pituitaries contributed to each recipient of the hormone, the process amplified the very few natural cases of CJD into a real epidemic. But if you condemn science for a Faustian meddling with nature that backfired, give it the credit for solving this problem, too. Even before the extent of the growth-hormone CJD epidemic had been recognised in 1984, synthetic growth hormone, one of the first products to come from genetically engineered bacteria, was replacing the cadaver-derived hormone.

Let us take stock of this strange tale as it appeared in about 1980. Sheep, mink, monkeys, mice and people could all acquire versions of the same disease by the injection of contaminated brain. The contamination survived almost all normal germ-killing procedures and remained wholly invisible to even the most powerful electron microscopes. Yet it was not contagious in everyday life, did not seem to pass through mother's milk, raised no immune response, stayed latent for sometimes more than twenty or thirty years and could be caught from tiny doses – though the likelihood of contracting the disease depended strongly on the size of the dose received. What could it be?

Almost forgotten in the excitement was the case of the Suffolk sheep and the hint that inbreeding had exacerbated scrapie at the outset. It was also gradually becoming clear that in a few human cases – though fewer than six per cent – there seemed to be a family connection that hinted at a genetic disease. The key to understanding scrapie lay not in the arsenal of the pathologist, but in that of the geneticist. Scrapie was in the genes. Nowhere was this more starkly underlined than in Israel. When Israeli scientists sought out CJD in their own country in the mid-1970s, they noticed a remarkable thing. Fully fourteen of the cases, or thirty times more than expected by chance, were among the small number of Jews who had immigrated to Israel from Libya. Immediate suspicion fell upon their diet, which included a special predilection for sheep's brains. But no. The true explanation was genetic: all affected people were part of a single dispersed pedigree. They are now known to share a single mutation, one that is also found in a few families of Slovakians, Chileans and German-Americans.

The world of scrapie is eery and exotic yet also vaguely familiar. At the same time that one group of scientists were being irresistably drawn to the conclusion that scrapie was in the genes, another had been entertaining a revolutionary, indeed heretical, idea that seemed at first to be heading in a contradictory direction. As early as 1967 somebody had suggested that the scrapie agent might have no DNA or RNA genes at all. It might be the only piece of life on the planet that did not use nucleic acid and had no genes of its own. Since Francis Crick had recently coined what he called, half-seriously, the 'central dogma of genetics' – that DNA makes RNA makes protein – the suggestion that there was a living thing with no DNA was about as welcome in biology as Luther's principles in Rome.

In 1982 a geneticist named Stanley Prusiner proposed a resolution of the apparent paradox between a DNA-less creature and a disease that moved through human DNA. Prusiner had discovered a chunk of protein that resisted digestion by normal protease enzymes and that was present in animals with scrapie-like diseases but not in healthy versions of the same species. It was comparatively straight-forward for him to work out the sequence of amino acids in this protein chunk, calculate the equivalent DNA sequence and search for such sequences in amongst the genes of mice and, later, people. Prusiner thus found the gene, called *PRP* (for protease-resistant protein) and nailed his heresy to the church door of science. His theory, gradually elaborated over the next few years, went like this. *PRP* is a normal gene in mice and people; it produces a normal protein. It is not the gene of a virus. But its product, known as a prion, is a protein with an unusual quality: it can suddenly change its shape into a tough and sticky form that resists all attempts to destroy it and that gathers together in aggregate lumps, disrupting the structure of the cell. All this would be unprecedented enough, but Prusiner proposed something even more exotic. He suggested that this new form of prion has the capacity to reshape normal prions into versions of itself. It does not alter the sequence – proteins, like genes, are made of long, digital sequences – but it does change the way they fold up.[1]

Prusiner's theory fell on stony ground. It failed entirely to explain some of the most basic features of scrapie and related diseases, in particular, the fact that the diseases came in different strains. As he puts it ruefully today, 'Such a hypothesis enjoyed little enthusiasm.' I vividly remember the scorn with which scrapie experts greeted the Prusiner theory when I asked them for their views for an article I was writing about this time. But gradually, as the facts came in, it seemed as if he might have guessed right. It eventually became clear that a mouse with no prion genes cannot catch any of these diseases, whereas a dose of misshapen prion is sufficient to give the diseases to another mouse: the disease is both caused by prions and trans-mitted by them. But although the Prusiner theory has since felled a large forest of ignorance – and Prusiner duly followed Gajdusek to Stockholm to collect a Nobel prize – large woods remain. Prions retain deep mysteries, the foremost of which is what on earth they exist for. The *PRP* gene is not only present in every mammal so far examined, but it varies very little in sequence, which implies that it is doing some important job. That job almost certainly concerns the brain, which is where the gene is switched on. It may involve copper, which the prion seems to be fond of. But – and here's the mystery – a mouse in which both copies of the gene have been deliberately knocked out since before birth is a perfectly normal mouse. It seems that whatever function the prion serves, the mouse can grow to do without it. We are still no nearer to knowing why we have this potentially lethal gene.[2]

Meanwhile we live just a mutation or two away from catching the disease from our own prion genes. In human beings the gene has 253 'words' of three letters each, though the first twenty-two and the last twenty-three are cut off the protein as soon as it is manufactured. In just four places, a change of word can lead to prion disease – but to four different manifestations of the disease. Changing the 102nd word from proline to leucine causes Gerstmann–Straüssler–Scheinker disease, an inherited version of the disease that takes a long time to kill. Changing the 200th word from glutamine to lysine causes the type of CJD typical of the Libyan Jews. Changing the 178th word from aspartic acid to aspara- gine causes typical CJD, unless the 129th word is also changed from valine to methionine, in which case possibly the most horrible of all the prion diseases results. This is a rare affliction, known as fatal familial insomnia, where death occurs after months of total insomnia. In this case, it is the thalamus (which is, among other things, the brain's sleep centre), which is eaten away by the disease. It seems that the different symptoms of different prion diseases result from the erosion of different parts of the brain.

In the decade since these facts first became clear, science has been at its most magnificent in probing further into the mysteries of this one gene. Experiments of almost mind-boggling ingenuity have poured out of Prusiner's and others' laboratories, revealing a story of extraordinary determinism and specificity. The 'bad' prion changes shape by refolding its central chunk (words 108–121). A mutation in this region that makes the shape-change more likely is fatal so early in the life of a mouse that prion disease strikes within weeks of birth. The mutations that we see, in the various pedigrees of inherited prion disease, are peripheral ones that only slightly change the odds of the change in shape. In this way science tells us more and more about prions, but each new piece of knowledge only exposes a greater depth of mystery.

How exactly is this shape change effected? Is there, as Prusiner suspects, an unidentified second protein involved, called protein X, and if so, why can we not find it? We do not know.

How can it be that the same gene, expressed in all parts of the brain, behaves differently in different parts of the brain depending on which mutation it has? In goats, the symptoms of the disease vary from sleepiness to hyperactivity depending on which of two strains of the disease they get. We do not know why this should be.

Why is there a species barrier, which makes it hard to transmit prion diseases between species, but easy within species? Why is it very difficult to transmit by the oral route, but comparatively easy by means of direct injection into the brain? We do not know.

Why is the onset of symptoms dose-dependent? The more prions a mouse ingests, the sooner it will show symptoms. The more copies of a prion gene that a mouse has, the more quickly it can get prion disease when injected with rogue prions. Why? We do not know.

Why is it safer to be heterozygous than homozygous? In other words, if you have, at word 129, a valine on one copy of the gene and a methionine on the other copy, why are you more resistant to prion diseases (except fatal familial insomnia) than somebody who has either two valines or two methionines? We do not know.

Why is the disease so picky? Mice cannot easily get hamster scrapie, nor vice versa. But a mouse deliberately equipped with a hamster prion gene will catch hamster scrapie from an injection of hamster brains. A mouse equipped with two different versions of human prion genes can catch two kinds of human disease, one like fatal familial insomnia and one like CJD. A mouse equipped with both human and mouse prion genes will be slower to get human CJD than a mouse with only a human prion gene: does this mean different prions compete? We do not know.

How does the gene change its strain as it moves through a new species? Mice cannnot easily catch hamster scrapie, but when they do, they pass it on with progressively greater ease to other mice.[3] Why? We do not know.

Why does the disease spread from the site of injection slowly and progressively, as if bad prions can only convert good prions in their immediate vicinity? We know the disease moves through the B cells of the immune system, which somehow transmit it to the brain.[4] But why them, and how? We do not know.

The truly baffling aspect of this proliferating knowledge of ignorance is that it strikes at the heart of an even more central genetic dogma than Francis Crick's. It undermines one of the messages I have been evangelising since the very first chapter of this book, that the core of biology is digital. Here, in the prion gene, we have respectable digital changes, substituting one word for another, yet causing changes that cannot be wholly predicted without other knowledge. The prion system is analogue, not digital. It is a change not of sequence but of shape and it depends on dose, location and whether the wind is in the west. That is not to say it lacks determinism. If anything, CJD is even more precise than Huntington's disease in the age at which it strikes. The record includes cases of siblings who caught it at exactly the same age despite living apart all their lives.

Prion diseases are caused by a sort of chain reaction in which one prion converts its neighbour to its own shape and they each then convert another, and so on, exponentially. It is just like the fateful image that Leo Szilard conjured in his brain one day in 1933, while waiting to cross a street in London: the image of an atom splitting and emitting two neutrons, which caused another atom to split and emit two neutrons, and so on – the image of the chain reaction that later exploded over Hiroshima. The prion chain reaction is of course much slower than the neutron one. But it is just as capable of exponential explosion; the New Guinea kuru epidemic stood as proof of this possibility even as Prusiner began to tease out the details in the early 1980s. Yet already, much closer to home, an even bigger epidemic of prion disease was just starting its chain reaction. This time the victims were cows.

Nobody knows exactly when, where or how – that damned mystery again – but at some time in the late 1970s or early 1980s the British manufacturers of processed cattle food began to incorporate misshapen prions into their product. It might have been because a change in the processes in rendering factories followed a fall in the price of tallow. It might have been because rising numbers of old sheep found their way into the factories thanks to generous lamb subsidies. Whichever was the cause, the wrong-shaped prions got into the system: all it took was one highly infectious animal, riddled with scrapified prions, rendered into cattle cake. No matter that the bones and offal from old cows and sheep were boiled to sterility as they were rendered into protein-rich supplements for dairy cattle. Scrapified prions can survive boiling.

The chances of giving a cow prion disease would still have been very small, but with hundreds of thousands of cows it was enough. As soon as the first cases of 'mad-cow disease' went back into the food chain to be made into feed for other cows, the chain reaction had begun. More and more prions came through into the cattle cake, giving larger and larger doses to new calves. The long incubation period meant that doomed animals took five years on average to show symptoms. When the first six cases were recognised as something unusual by the end of 1986, there were already roughly 50,000 doomed animals in Britain, though nobody could possibly have known it. Eventually about 180,000 cattle died of bovine spongiform encephalopathy (BSE) before the disease was almost eradicated in the late 1990s.

Within a year of the first reported case, skilful detective work by government vets had pinned down the source of the problem as contaminated feed. It was the only theory that fitted all the details and it accounted for strange anomalies such as the fact that the island of Guernsey had an epidemic long before Jersey: the two islands had two different feed suppliers, one of which used much meat and bonemeal, while the other used little. By July 1988 the Ruminant Feed Ban was law. It is hard to see how experts or ministers could have acted more quickly, except with perfect hindsight. By August 1988, the Southwood committee's recommendation that all BSE-infected cattle be destroyed and not allowed to enter the food chain had been enacted. This was when the first blunder was made: the decision to pay only fifty per cent of each animal's value in compensation, thus providing an incentive to farmers to ignore signs of the disease. But even this mistake may not have been as costly as people assume: when compensation was increased, there was no jump in the number of cases notified.

The Specified Bovine Offals Ban, preventing adult cows' brains from entering the human food chain, came into force a year later, and was only extended to calves in 1990. This might have happened sooner, but, given what was known about the difficulty other species had of catching sheep scrapie except by direct injection of brain into brain, at the time it seemed too cautious. It had proved impossible to infect monkeys with human prion diseases through their food, except by using huge doses: and the jump from cow to person is a much bigger jump than from person to monkey. (It has been estimated that intracerebral injection magnifies the risk 100 million times compared with ingestion.) To say anything other than that beef was 'safe' to eat at this stage would have been the height of irresponsibility.

As far as scientists were concerned, the risk of cross-species transmission by the oral route was indeed vanishingly small: so small that it would be impossible to achieve a single case in an experiment without hundreds of thousands of experimental animals. But that was the point: the experiment was being conducted with fifty million experimental animals called Britons. In such a large sample, a few cases were inevitable. For the politician, safety was an absolute, not a relative matter. They did not want few human cases; they wanted no human cases. Besides, BSE, like every prion disease before it, was proving alarmingly good at springing surprises. Cats were catching it from the same meat and bonemeal that cattle ate – more than seventy domestic cats, plus three cheetahs, a puma, an ocelot and even a tiger have since died of BSE. But no case of dog BSE has yet appeared. Were people going to be as resistant as dogs or as susceptible as cats?

By 1992, the cattle problem was effectively solved, although the peak of the epidemic was still to come because of the five-year lag between infection and symptoms. Very few cattle born since 1992 have caught or will catch BSE. Yet the human hysteria was only just beginning. It was now that the decisions taken by politicians started to grow steadily more lunatic. Thanks to the offals ban, beef was now safer to eat than at any time in ten years, yet it was only now that people began to boycott it.

In March 1996, the government announced that ten people had indeed died of a form of prion disease that looked suspiciously as if it had been transmitted from beef during the dangerous period: it resembled BSE in some symptoms and it had never been seen before. Public alarm, fanned by a willing press, became – briefly – extreme. Wild predictions of millions of deaths in Britain alone were taken seriously. The folly of turning cattle into cannibals was widely portrayed as an argument for organic farming. Conspiracy theories abounded: that the disease was caused by pesticides; that scientists were being muzzled by politicians; that the true facts were being suppressed; that deregulation of the feed industry had caused the problem; that France, Ireland, Germany and other countries were suppressing news of epidemics just as large.

The government felt obliged to respond with a further useless ban, on the consumption of any cow over thirty months of age: a ban that further inflamed public alarm, ruined a whole industry and choked the system with doomed cattle. Later that year, at the insistence of European politicians, the government ordered the 'selective cull' of 100,000 more cattle, even though it knew this was a meaningless gesture that would further alienate farmers and consumers. It was no longer even shutting the stable door after the horse had bolted; it was sacrificing a goat outside the stable. Predictably, the new cull did not even have the effect of lifting the European Union's largely self-interested ban on all British beef exports. But worse was to follow with the ban on beef on the bone in 1997. Everybody agreed that the risk from beef on the bone was infinitesimal – likely to lead to at most one case of CJD every four years. The government's approach to risk was now so nationalising that the agriculture minister was not even prepared to let people make up their own minds about a risk smaller than that of being struck by lightning. By taking such an absurd attitude to risk, indeed, the government predictably provoked riskier behaviour in its subjects. In some circles, almost a mood of civil disobedience obtained, and I found myself offered more oxtail stew as the ban loomed than I had ever done before.

Throughout 1996, Britain braced itself for an epidemic of human BSE. Yet in the year from March only six people died of the disease. Far from growing, the numbers seemed to be steady or falling. As I write, it is still uncertain how many people will die of 'new-variant' CJD. The figure has inched up past forty, each case an almost unimaginable family tragedy, but not yet an epidemic. At first, the victims of this new-variant CJD appeared, on investigation, to be particularly enthusiastic meat-eaters in the dangerous years, even though one of the first cases had turned vegetarian some years before. But this was an illusion: when scientists asked the relatives of those thought to have died of CJD (but who, *post mortem*, were proved to have died of something else) about their habits, they found the same meat-eating bias: the memories said more about the psychology of the relatives than reality.

The one thing the victims had in common was that almost all were of one genotype–homozygous for methionine at 'word' 129. Perhaps the far more numerous heterozygotes and valine-homozygotes will prove simply to have a longer incubation period: BSE transmitted to monkeys by intracerebral injection has a much longer incubation period than most prion diseases. On the other hand, given that the vast majority of human infections from beef would have occurred before the end of 1988, and ten years is already twice as long as the average incubation period in cattle, maybe the species barrier is as high as it seems in animal experiments and we have already seen the worst of the epidemic. Maybe, too, the new-variant CJD has nothing to do with beef-eating. Many now believe that the possibility that human vaccines and other medical products, made with beef products, posed a much greater danger was somewhat too hastily rejected by the authorities in the late 1980s.

CJD has killed lifelong vegetarians who had never had surgery, never left Britain and never worked on a farm or in a butcher's shop. The last and greatest mystery of the prion is that even today – when forms of CJD have been caught by all sorts of known means, including cannibalism, surgery, hormone injections and possibly beef-eating – eighty-five per cent of all CJD cases are 'sporadic', meaning that they cannot at the moment be explained by anything other than random chance. This offends our natural determinism, in which diseases must have causes, but we do not live in a fully determined world. Perhaps CJD just happens spontaneously at the rate of about one case per million people.

Prions have humbled us with our ignorance. We did not suspect that there was a form of self-replication that did not use DNA – did not indeed use digital information at all. We did not imagine that a disease of such profound mystery could emerge from such unlikely quarters and prove so deadly. We still do not quite see how changes in the folding of a peptide chain can cause such havoc, or how tiny changes in the composition of the chain can have such complicated implications. As two prion experts have written,[5] 'Personal and family tragedies, ethnological catastrophes and economic disasters can all be traced back to the mischievous misfolding of one small molecule.'

Eugenics

I know no safe depository of the ultimate powers of the
society but the people themselves, and if we think them
not enlightened enough to exercise that control with a
wholesome discretion, the remedy is not to take it from
them, but to inform their discretion.   *Thomas Jefferson*

Chromosome 21 is the smallest human chromosome. It ought, as
a result, to be called chromosome 22, but the chromosome that has
that name was until recently thought to be smaller still and the name
is now established. Perhaps because it is the smallest chromosome,
with probably the fewest genes, chromosome 21 is the only chromosome that can be present in three copies rather than two in a healthy
human body. In all other cases, having an extra copy of a whole
chromosome so upsets the balance of the human genome that the
body cannot properly develop at all. Children are occasionally born
with an extra chromosome 13 or 18, but they never survive more
than a few days. Children born with an extra chromosome 21 are
healthy, conspicuously happy and destined to live for many years.
But they are not considered, in that pejorative word, 'normal'. They

have Down syndrome. Their characteristic appearance – short stature, plump bodies, narrow eyes, happy faces – is immediately
familiar. So is the fact that they are mentally retarded, gentle and
destined to age rapidly, often developing a form of Alzheimer's
disease, and die before they reach the age of forty.

Down-syndrome babies are generally born to older mothers. The
probability of having a Down-syndrome baby grows rapidly and
exponentially as the age of the mother increases, from 1 in 2,300 at
the age of twenty to 1 in 100 at forty. It is for this reason alone
that Down embryos are the principal victims, or their mothers the
principal users, of genetic screening. In most countries amniocentesis
is now offered to – perhaps even imposed on – all older mothers,
to check whether the foetus carries an extra chromosome. If it does,
the mother is offered – or cajoled into – an abortion. The reason
given is that, despite the happy demeanour of these children, most
people would rather not be the parent of a Down child. If you are
of one opinion, you see this as a manifestation of benign science,
miraculously preventing the birth of cruelly incapacitated people at
no suffering. If you are of another opinion you see the officially
encouraged murder of a sacred human life in the dubious name of
human perfection and to the disrespect of disability. You see, in
effect, eugenics still in action, more than fifty years after it was
grotesquely discredited by Nazi atrocities.

This chapter is about the dark side of genetics' past, the black
sheep of the genetics family – the murder, sterilisation and abortion
committed in the name of genetic purity.

The father of eugenics, Francis Galton, was in many ways the
opposite of his first cousin, Charles Darwin. Where Darwin was
methodical, patient, shy and conventional, Galton was an intellectual
dilettante, a psychosexual mess and showman. He was also brilliant.
He explored southern Africa, studied twins, collected statistics and
dreamed of Utopias. Today his fame is almost as great as his cousin's,
though it is something more like notoriety than fame. Darwinism
was always in danger of being turned into a political creed and
Galton did so. The philosopher Herbert Spencer had enthusiastically

embraced the idea of survival of the fittest, arguing that it buttressed the credibility of laissez-faire economics and justified the individualism of Victorian society: social darwinism, he called it. Galton's vision was more prosaic. If, as Darwin had argued, species had been altered by systematic selective breeding, like cattle and racing pigeons, then so could human beings be bred to improve the race. In a sense Galton appealed to an older tradition than Darwinism: the eighteenth-century tradition of cattle breeding and the even older breeding of apple and corn varieties. His cry was: let us improve the stock of our own species as we have improved that of others. Let us breed from the best and not from the worst specimens of humanity. In 1885 he coined the term 'eugenic' for such breeding.

But who was 'us'? In a Spencerian world of individualism, it was literally each one of us: eugenics meant that each individual strove to pick a good mate – somebody with a good mind and a healthy body. It was little more than being selective about our marriage partners – which we already were. In the Galtonian world, though, 'us' came to mean something more collective. Galton's first and most influential follower was Karl Pearson, a radical socialist utopian and a brilliant statistician. Fascinated and frightened by the growing economic power of Germany, Pearson turned eugenics into a strand of jingoism. It was not the individual that must be eugenic; it was the nation. Only by selectively breeding among its citizens would Britain stay ahead of its continental rival. The state must have a say in who should breed and who should not. At its birth eugenics was not a politicised science; it was a science-ised political creed.

By 1900, eugenics had caught the popular imagination. The name Eugene was suddenly in vogue and there was a groundswell of popular fascination with the idea of planned breeding, as eugenics meetings popped up all over Britain. Pearson wrote to Galton in 1907: 'I hear most respectable middle-class matrons saying, if children are weakly, "Ah, but that was not a eugenic marriage!"' The poor condition of Boer War recruits to the army stimulated as much debate about better breeding as it did about better welfare.

Something similar was happening in Germany, where a mixture

of Friedrich Nietzsche's philosophy of the hero and Ernst Haeckel's doctrine of biological destiny produced an enthusiasm for evolutionary progress to go with economic and social progress. The easy gravitation to an authoritarian philosophy meant that in Germany, even more than in Britain, biology became enmeshed in nationalism. But for the moment it remained largely ideological, not practical.[1]

So far, so benign. The focus soon shifted, however, from encouraging the 'eugenic' breeding of the best to halting the 'dysgenic' breeding of the worst. And the 'worst' soon came to mean mainly the 'feeble-minded', which included alcoholics, epileptics and criminals as well as the mentally retarded. This was especially true in the United States, where in 1904 Charles Davenport, an admirer of Galton and Pearson, persuaded Andrew Carnegie to found for him the Cold Spring Harbor Laboratory to study eugenics. Davenport, a strait-laced conservative with immense energy, was more concerned with preventing dysgenic breeding than urging eugenic breeding. His science was simplistic to say the least; for example, he said that now that Mendelism had proved the particulate nature of inheritance, the American idea of a national 'melting pot' could be consigned to the past; he also suggested that a naval family had a gene for thalassophilia, or love of the sea. But in politics, Davenport was skilled and influential. Helped along by a successful book by Henry Goddard about a largely mythical, mentally deficient family called the Kallikaks, in which the case was strongly made that feeble-mindedness was inherited, Davenport and his allies gradually persuaded American political opinion that the race was in desperate danger of degeneracy. Said Theodore Roosevelt: 'Some day we will realise that the prime duty, the inescapable duty, of the good citizen of the right type is to leave his or her blood behind him in the world.' Wrong types need not apply.[2]

Much of the American enthusiasm for eugenics stemmed from anti-immigrant feeling. At a time of rapid immigration from eastern and southern Europe, it was easy to whip up paranoia that the 'better' Anglo-Saxon stock of the country was being diluted. Eugenic arguments provided a convenient cover for those who wished to restrict immigration for more traditional, racist reasons. The Immigration Restriction Act of 1924 was a direct result of eugenic campaigning. For the next twenty years it consigned many desperate European emigrants to a worse fate at home by denying them a new home in the United States, and it remained on the books unamended for forty years.

Restricting immigration was not the only legal success for the eugenists. By 1911 six states already had laws on their books to allow the forced sterilisation of the mentally unfit. Six years later another nine states had joined them. If the state could take the life of a criminal, so went the argument, then surely it could deny the right to reproduce (as if mental innocence were on a par with criminal guilt). 'It is the acme of stupidity ... to talk in such cases of individual liberty, or the rights of the individual. Such individuals ... have no right to propagate their kind.' So wrote an American doctor named W. J. Robinson.

The Supreme Court threw out many sterilisation laws at first, but in 1927 it changed its line. In *Buck v. Bell*, the court ruled that the commonwealth of Virginia could sterilise Carrie Buck, a seventeen-year-old girl committed to a colony for epileptics and the feeble minded in Lynchburg, where she lived with her mother Emma and her daughter Vivian. After a cursory examination, Vivian, who was seven months old (!), was declared an imbecile and Carrie was ordered to be sterilised. As Justice Oliver Wendell Holmes famously put it in his judgment, 'Three generations of imbeciles are enough.' Vivian died young, but Carrie survived into old age, a respectable woman of moderate intelligence who did crossword puzzles in her spare time. Her sister Doris, also sterilised, tried for many years to have babies before realising what had been done to her without her consent. Virginia continued to sterilise the mentally handicapped into the 1970s. America, bastion of individual liberty, sterilised more than 100,000 people for feeble-mindedness, under more than 30 state and federal laws passed between 1910 and 1935.

But although America was the pioneer, other countries followed. Sweden sterilised 60,000. Canada, Norway, Finland, Estonia and

Iceland all put coercive sterilisation laws on their books and used them. Germany, most notoriously, first sterilised 400,000 people and then murdered many of them. In just eighteen months in the Second World War, 70,000 already-sterilised German psychiatric patients were gassed just to free hospital beds for wounded soldiers.

But Britain, almost alone among Protestant industrial countries, never passed a eugenic law: that is, it never passed a law allowing the government to interfere in the individual's right to breed. In particular, there was never a British law preventing marriage of the mentally deficient, and there was never a British law allowing compulsory sterilisation by the state on the grounds of feeble-mindedness. (This is not to deny that there has been individual 'freelance' practice of cajoled sterilisation by doctors or hospitals.)

Britain was not unique; in countries where the influence of the Roman Catholic church was strong, there were no eugenic laws. The Netherlands avoided passing such laws. The Soviet Union, more concerned about purging and killing clever people than dull ones, never put such a law on its books. But Britain stands out, because it was the source of much – indeed most – eugenic science and propaganda in the first forty years of the twentieth century. Rather than ask how so many countries could have followed such cruel practices, it is instructive to turn the question on its head: why did Britain resist the temptation? Who deserves the credit?

Not the scientists. Scientists like to tell themselves today that eugenics was always seen as a 'pseudoscience' and frowned on by true scientists, especially after the rediscovery of Mendelism (which reveals how many more silent carriers of mutations there are than frank mutants), but there is little in the written record to support this. Most scientists welcomed the flattery of being treated as experts in a new technocracy. They were perpetually urging immediate action by government. (In Germany, more than half of all academic biologists joined the Nazi party – a higher proportion than in any other professional group – and not one criticised eugenics.[3])

A case in point is Sir Ronald Fisher, yet another founder of modern statistics (although Galton, Pearson and Fisher were great statisticians, nobody has concluded that statistics is as dangerous as genetics). Fisher was a true Mendelian, but he was also vice president of the Eugenics Society. He was obsessed with what he called 'the redistribution of the incidence of reproduction' from the upper classes to the poor: the fact that poor people had more children than rich people. Even later critics of eugenics like Julian Huxley and J. B. S. Haldane were supporters before 1920; it was the crudity and bias with which eugenic policies came to be adopted in the United States that they complained about, not the principle.

Nor could the socialists claim credit for stopping eugenics. Although the Labour party opposed eugenics by the 1930s, the socialist movement in general provided much of the intellectual ammunition for the movement before that. You have to dig hard to find a prominent British socialist in the first thirty years of the century who expressed even faint opposition to eugenic policies. It is extraordinarily easy to find pro-eugenic quotes from Fabians of the day. H. G. Wells, J. M. Keynes, George Bernard Shaw, Havelock Ellis, Harold Laski, Sidney and Beatrice Webb – all said creepy things about the urgent need to stop stupid or disabled people from breeding. A character in Shaw's *Man and superman* says: 'Being cowards, we defeat natural selection under cover of philanthropy: being sluggards, we neglect artificial selection under cover of delicacy and morality.'

The works of H. G. Wells are especially rich in juicy quotes: 'The children people bring into the world can be no more their private concern entirely than the disease germs they disseminate or the noises a man makes in a thin-floored flat.' Or 'The swarms of black, and brown, and dirty white, and yellow people . . . will have to go.' Or 'It has become apparent that whole masses of human population are, as a whole, inferior in their claim upon the future . . . to give them equality is to sink to their level, to protect and cherish them is to be swamped in their fecundity.' He added, reassuringly, 'All such killing will be done with an opiate.' (It wasn't.)[4]

Socialists, with their belief in planning and their readiness to put the state in a position of power over the individual, were ready-made for the eugenic message. Breeding, too, was ripe for nationalisation. It was among Pearson's friends in the Fabian Society that eugenics first took root as a popular theme. Eugenics was grist to the mill of their socialism. Eugenics was a progressive philosophy, and called for a role for the state.

Soon the Conservatives and Liberals were just as enthusiastic. Arthur Balfour, ex-prime minister, chaired the first International Eugenics Conference in London in 1912 and the sponsoring vice-presidents included the Lord Chief Justice and Winston Churchill. The Oxford Union approved the principles of eugenics by nearly two to one in 1911. As Churchill put it, 'the multiplication of the feeble-minded' was 'a very terrible danger to the race'.

To be sure, there were a few lone voices of dissent. One or two intellectuals remained suspicious, among them Hilaire Belloc and G. K. Chesterton, who wrote that 'eugenicists had discovered how to combine hardening of the heart with softening of the head'. But be in no doubt that most Britons were in favour of eugenic laws.

There were two moments when Britain very nearly did pass eugenic laws: in 1913 and 1934. In the first case, the attempt was thwarted by brave and often lonely opponents swimming against the tide of conventional wisdom. In 1904 the government set up a Royal Commission under the Earl of Radnor on the 'care and control of the feeble-minded'. When it reported in 1908, it took a strongly hereditarian view of mental deficiency, which was not surprising given that many of its members were paid-up eugenists. As Gerry Anderson has demonstrated in a recent Cambridge thesis,[5] there followed a period of sustained lobbying by pressure groups to try to persuade the government to act. The Home Office received hundreds of resolutions from county and borough councils and from education committees, urging the passage of a bill that would restrict reproduction by the 'unfit'. The new Eugenics Education Society bombarded MPs and had meetings with the Home Secretary to further the cause.

For a while nothing happened. The Home Secretary, Herbert Gladstone, was unsympathetic. But when he was replaced by

Winston Churchill in 1910, eugenics at last had an ardent champion at the cabinet table. Churchill had already in 1909 circulated as a cabinet paper a pro-eugenics speech by Alfred Tredgold. In December 1910, installed in the Home Office, Churchill wrote to the Prime Minister, Herbert Asquith, advocating urgent eugenic legislation and concluding: 'I feel that the source from which the stream of madness is fed should be cut off and sealed up before another year has passed.' He wanted for mental patients that their 'curse would die with them'. In case there is any doubt of what he meant, Wilfred Scawen Blunt wrote that Churchill was already privately advocating the use of X-rays and operations to sterilise the mentally unfit.

The constitutional crises of 1910 and 1911 prevented Churchill introducing a bill and he moved on to the Admiralty. But by 1912 the clamour for legislation had revived and a Tory backbencher, Gershom Stewart, eventually forced the government's hand by introducing his own private member's bill on the matter. In 1912 the new Home Secretary, Reginald McKenna, somewhat reluctantly brought in a government bill, the Mental Deficiency Bill. The bill would restrict procreation by feebled-minded people and would punish those who married mental defectives. It was an open secret that it could be amended to allow compulsory sterilisation as soon as practicable.

One man deserves to be singled out for mounting opposition to this bill: a radical libertarian MP with the famous – indeed relevant – name of Josiah Wedgwood. Scion of the famous industrial family that had repeatedly intermarried with the Darwin family – Charles Darwin had a grandfather, a father-in-law and a brother-in-law (twice over) each called Josiah Wedgwood – the latest Josiah was a naval architect by profession. He had been elected to Parliament in the Liberal landslide of 1906, but later joined the Labour party and retired to the House of Lords in 1942. (Darwin's son, Leonard, was at the time president of the Eugenics Society.)

Wedgwood disliked eugenics intensely. He charged that the Eugenics Society was trying 'to breed up the working class as though they were cattle' and he asserted that the laws of heredity were 'too undetermined for one to pin faith on any doctrine, much less to legislate according to it'. But his main objection was on the grounds of individual liberty. He was appalled at a bill that gave the state powers to take a child from its own home by force, by clauses that granted policemen the duty to act upon reports from members of the public that somebody was 'feeble-minded'. His motive was not social justice, but individual liberty: he was joined by Tory libertarians such as Lord Robert Cecil. Their common cause was that of the individual against the state.

The clause that really stuck in Wedgwood's throat was the one that stated it to be 'desirable in the interests of the community that [the feeble-minded] should be deprived of the opportunity of procreating children.' This was, in Wedgwood's words, 'the most abominable thing ever suggested' and not 'the care for the liberty of the subject and for the protection of the individual against the state that we have a right to expect from a Liberal Administration'.[6]

Wedgwood's attack was so effective that the government withdrew the bill and presented it again the next year in much watered-down form. Crucially, it now omitted 'any reference to what might be regarded as the eugenic idea' (in McKenna's words), and the offensive clauses regulating marriage and preventing procreation were dropped. Wedgwood still opposed the bill and for two whole nights, fuelled by bars of chocolate, he sustained his attack by tabling more than 200 amendments. But when his support had dwindled to four members, he gave up and the bill passed into law.

Wedgwood probably thought he had failed. The forcible committal of mental patients became a feature of British life and this in practice did make it harder for them to breed. But in truth he had not only prevented eugenic measures being adopted; he had also sent a warning shot across the bows of any future government that eugenic legislation could be contentious. And he had identified the central flaw in the whole eugenic project. This was not that it was based on faulty science, nor that it was impractical, but that it was fundamentally oppressive and cruel, because it required the full power of the state to be asserted over the rights of the individual.

In the early 1930s, as unemployment rose during the depression, eugenics experienced a marked revival. In Britain the membership of eugenic societies reached record levels, as people began, absurdly, to blame high unemployment and poverty on the very racial degeneration that had been predicted by the first eugenists. It was now that most countries passed their eugenic laws. Sweden, for instance, implemented its compulsory sterilisation law in 1934, as did Germany.

Pressure for a British sterilisation law had again been building for some years, aided by a government report on mental deficiency known as the Wood report, which concluded that mental problems were on the increase and that this was partly due to the high fertility of mental defectives (this was the committee which carefully defined three categories of mental defectives: idiots, imbeciles and the feeble-minded). But when a private member's eugenics bill, introduced to the House of Commons by a Labour MP, was blocked, the eugenics pressure group changed tack and turned its attention to the civil service. The Department of Health was persuaded to appoint a committee under Sir Laurence Brock to examine the case for sterilising the mentally unfit.

The Brock committee, despite its bureaucratic origins, was partisan from the outset. Most of its members were, according to a modern historian, 'not even in a weak sense actuated by a desire to consider dispassionately the contradictory and inconclusive evidence'. The committee accepted a hereditarian view of mental deficiency, ignoring the evidence against this and 'padding' (its word) the evidence in favour. It accepted the notion of a fast-breeding mental underclass, despite inconclusive evidence, and it 'rejected' compulsory sterilisation only to help assuage critics – it glossed over the problem of obtaining consent from mentally defective people. A quotation from a popular book about biology published in 1931 gives the game away: 'Many of these low types might be bribed or otherwise persuaded to accept voluntary sterilization.'[7]

The Brock report was the purest propaganda, dressed up as a dispassionate and expert assessment of the issues. As has been pointed out recently, in the way it created a synthetic crisis, endorsed by a consensus of 'experts' and requiring urgent action, it was a harbinger of the way international civil servants would behave much later in the century over global warming.[8]

The report was intended to lead to a sterilisation bill, but such a bill never saw the light of day. This time it was not so much because of a determined contrarian like Wedgwood, but because of a changing climate of opinion throughout society. Many scientists had changed their minds, notably J. B. S. Haldane, partly because of the growing influence of environmental explanations of human nature promulgated by people like Margaret Mead and the behaviourists in psychology. The Labour party was now firmly against eugenics, which it saw as a form of class war on the working class. The opposition of the Catholic Church was also influential in some quarters.[9]

Surprisingly, it was not until 1938 that reports filtered through from Germany of what compulsory sterilisation meant in practice. The Brock committee had been unwise enough to praise the Nazi sterilisation law, which came into force in January 1934. It was now clear that this law was an intolerable infringement of personal liberty and an excuse for persecution. In Britain, good sense prevailed.[10]

This brief history of eugenics leads me to one firm conclusion. What is wrong with eugenics is not the science, but the coercion. Eugenics is like any other programme that puts the social benefit before the individual's rights. It is a humanitarian, not a scientific crime. There is little doubt that eugenic breeding would 'work' for human beings just as it works for dogs and dairy cattle. It would be possible to reduce the incidence of many mental disorders and improve the health of the population by selective breeding. But there is also little doubt that it could only be done very slowly at a gigantic cost in cruelty, injustice and oppression. Karl Pearson once said, in answer to Wedgwood: 'What is social is right, and there is no definition of right beyond that.' That dreadful statement should be the epitaph of eugenics.

Yet, as we read in our newspapers of genes for intelligence, of germline gene therapy, of prenatal diagnosis and screening, we cannot but feel in our bones that eugenics is not dead. As I argued in the chapter on chromosome 6, Galton's conviction that much of human nature has a hereditary element is back in fashion, this time with better – though not conclusive – empirical evidence. Increasingly, today, genetic screening allows parents to choose the genes of their children. The philosopher Philip Kitcher, for instance, calls genetic screening 'laissez-faire eugenics': 'Everyone is to be his (or her) own eugenicist, taking advantage of the available genetic tests to make the reproductive decisions she (he) thinks correct.'[11]

By this standard, eugenics happens every day in hospitals all over the world and by far its most common victims are embryos equipped with an extra chromosome 21, who would otherwise be born with Down syndrome. In most cases, had they been born, they would have led short, but largely happy lives – that is the nature of their disposition. In most cases, had they been born, they would have been loved by parents and siblings. But for a dependent, non-sentient embryo, not being born is not necessarily the same as being killed. We are back, in short order, to the debate on abortion and whether the mother has the right to abort a child, or the state the right to stop her: an old debate. Genetic knowledge gives her more reasons for wanting an abortion. The possibility of choosing among embryos for special ability, rather than against lack of ability, may not be too far away. Choosing boys and aborting girls is already a rampant abuse of amniocentesis in the Indian subcontinent in particular.

Have we rejected government eugenics merely to fall into the trap of allowing private eugenics? Parents may come under all sorts of pressures to adopt voluntary eugenics, from doctors, from health-insurance companies and from the culture at large. Stories abound of women as late as the 1970s being cajoled by their doctors into sterilisation because they carried a gene for a genetic disease. Yet if government were to ban genetic screening on the grounds that it might be abused, it would risk increasing the load of suffering in the world: it would be just as cruel to outlaw screening as to make it compulsory. It is an individual decision, not one that can be left to technocrats. Kitcher certainly thinks so: 'As for the traits that people attempt to promote or avoid, that is surely their own business.' So does James Watson: 'These things should be kept away from people who think they know best . . . I am trying to see genetic decisions put in the hand of users, which governments aren't.'[12]

Although there are still a few fringe scientists worried about the genetic deterioration of races and populations,[13] most scientists now recognise that the well-being of individuals should take priority over that of groups. There is a world of difference between genetic screening and what the eugenists wanted in their heyday – and it lies in this: genetic screening is about giving private individuals private choices on private criteria. Eugenics was about nationalising that decision to make people breed not for themselves but for the state. It is a distinction frequently overlooked in the rush to define what 'we' must allow in the new genetic world. Who is 'we'? We as individuals, or we as the collective interest of the state or the race?

Compare two modern examples of 'eugenics' as actually practised today. In the United States, as I discussed in the chapter on chromosome 13, the Committee for the Prevention of Jewish Genetic Disease tests schoolchildren's blood and advises against later marriages in which both parties carry the same disease-causing version of a particular gene. This is an entirely voluntary policy. Although it has been criticised as eugenic, there is no coercion involved at all.[14]

The other example comes from China, where the government continues to sterilise and abort on eugenic grounds. Chen Mingzhang, minister of public health, recently expostulated that births of inferior quality are serious among 'the old revolutionary base, ethnic minorities, the frontier, and economically poor areas'. The Maternal and Infant Health Care Law, which came into effect only in 1994, makes premarital check-ups compulsory and gives to doctors, not parents, the decision to abort a child. Nearly ninety per cent of Chinese geneticists approve of this compared with five per cent of American geneticists; by contrast eighty-five per cent of the American geneticists think an abortion decision should be made by the woman, compared with forty-four per cent of the Chinese. As

Xin Mao, who conducted the Chinese part of this poll, put it, echoing Karl Pearson: 'The Chinese culture is quite different, and things are focused on the good of society, not the good of the individual.'[15]

Many modern accounts of the history of eugenics present it as an example of the dangers of letting science, genetics especially, out of control. It is much more an example of the danger of letting government out of control.

CHROMOSOME 22

◆◆

Free Will

Hume's fork: Either our actions are determined, in which case we are not responsible for them, or they are the result of random events, in which case we are not responsible for them.

Oxford Dictionary of Philosophy

As this book is being completed, a few months before the end of a millennium, there comes news of a momentous announcement. At the Sanger Centre, near Cambridge – the laboratory which leads the world in reading the human genome – the complete sequence of chromosome 22 is finished. All 15.5 million 'words' (or so – the exact length depends on the repeat sequences, which vary greatly) in the twenty-second chapter of the human autobiography have been read and written down in English letters: 47 million As, Cs, Gs and Ts.

Near the tip of the long arm of chromosome 22 there lies a massive and complicated gene, pregnant with significance, known as *HFW*. It has fourteen exons, which together spell out a text more than 6,000 letters long. That text is severely edited after transcription by the strange process of RNA splicing to produce a highly complicated protein that is expressed only in a small part of the prefrontal cortex of the brain. The function of the protein is, generalising horribly, to endow human beings with free will. Without *HFW*, we would have no free will.

The preceding paragraph is fictional. There is no *HFW* gene on chromosome 22 nor on any other. After twenty-two chapters of relentless truth, I just felt like deceiving you. I cracked under the strain of being a non-fiction writer and could no longer resist the temptation to make something up.

But who am 'I'? The I who, overcome by a silly impulse, decided to write a fictional paragraph? I am a biological creature put together by my genes. They prescribed my shape, gave me five fingers on each hand and thirty-two teeth in my mouth, laid down my capacity for language, and defined about half of my intellectual capacity. When I remember something, it is they that do it for me, switching on the CREB system to store the memory. They built me a brain and delegated responsibility for day-to-day duties to it. They also gave me the distinct impression that I am free to make up my own mind about how to behave. Simple introspection tells me there is nothing that I 'cannot help myself' doing. There is equally nothing that says that I must do one thing and not something else. I am quite capable of jumping in my car and driving to Edinburgh right now and for no other reason than that I want to, or of making up a whole paragraph of fiction. I am a free agent, equipped with free will.

Where did this free will come from? It plainly could not have come from my genes, or else it would not be free will. The answer, according to many, is that it came from society, culture and nurture. According to this reasoning, freedom equals the parts of our natures not determined by our genes, a sort of flower that blooms after our genes have done their tyrannical worst. We can rise above our genetic determinism and grasp that mystic flower, freedom.

There has been a long tradition among a certain kind of science writer to say that the world of biology is divided into people who believe in genetic determinism and people who believe in freedom. Yet these same writers have rejected genetic determinism only by establishing other forms of biological determinism in its place – the determinism of parental influence or social conditioning. It is odd that so many writers who defend human dignity against the tyranny of our genes seem happy to accept the tyranny of our surroundings. I was once criticised in print for allegedly saying (which I had not) that all behaviour is genetically determined. The writer went on to give an example of how behaviour was not genetic: it was well known that child abusers were generally abused themselves as children and this was the cause of their later behaviour. It did not seem to occur to him that this was just as deterministic and a far more heartless and prejudicial condemnation of people who had suffered enough than anything I had said. He was arguing that the children of child abusers were likely to become child abusers and there was little they could do about it. It did not occur to him that he was applying a double standard: demanding rigorous proof for genetic explanations of behaviour while easily accepting social ones.

The crude distinction between genes as implacable programmers of a Calvinist predestination and the environment as the home of liberal free will is a fallacy. One of the most powerful environmental sculptors of character and ability is the sum of conditions in the womb, about which you can do nothing. As I argued in the chapter on chromosome 6, some of the genes for intellectual ability are probably genes for appetite rather than aptitude: they set their possessor on a course of willing learning. The same result can be achieved by an inspiring teacher. Nature, in other words, can be much more malleable than nurture.

Aldous Huxley's *Brave new world*, written at the height of eugenic enthusiasm in the 1920s, presents a terrifying world of uniform, coerced control in which there is no individuality. Each person meekly and willingly accepts his or her place in a caste system – alphas to epsilons – and obediently does the tasks and enjoys the recreations that society expects of him or her. The very phrase ' brave new world' has come to mean such a dystopia brought into being by central control and advanced science working hand-in-hand.

It therefore comes as something of a surprise to read the book and discover that there is virtually nothing about eugenics in it. Alphas and epsilons are not bred, but are produced by chemical adjustment in artificial wombs followed by Pavlovian conditioning and brainwashing, then sustained in adulthood by opiate-like drugs. In other words, this dystopia owes nothing to nature and everything to nurture. It is an environmental, not a genetic, hell. Everybody's fate is determined, but by their controlled environment, not their genes. It is indeed biological determinism, but not genetic determinism. Aldous Huxley's genius was to recognise how hellish a world in which nurture prevailed would actually be. Indeed, it is hard to tell whether the extreme genetic determinists who ruled Germany in the 1930s caused more suffering than the extreme environmental determinists who ruled Russia at the same time. All we can be sure of is that both extremes were horrible.

Fortunately we are spectacularly resistant to brainwashing. No matter how hard their parents or their politicians tell them that smoking is bad for them, young people still take it up. Indeed, it is precisely because grown-ups lecture them about it that it seems so appealing. We are genetically endowed with a tendency to be bloody-minded towards authority, especially in our teens, to guard our own innate character against dictators, teachers, abusing step-parents or government advertising campaigns.

Besides, we now know that virtually all the evidence purporting to show how parental influences shape our character is deeply flawed. There is indeed a correlation between abusing children and having been abused as a child, but it can be entirely accounted for by inherited personality traits. The children of abusers inherit their persecutor's characteristics. Properly controlled for this effect, studies leave no room for nurture determinism at all. The step-children of abusers, for instance, do not become abusers.[1]

The same, remarkably, is true of virtually every standard social nostrum you have ever heard. Criminals rear criminals. Divorcees rear divorcers. Problem parents rear problem children. Obese parents rear obese children. Having subscribed to all of these assertions during a long career of writing psychology textbooks, Judith Rich Harris suddenly began questioning them a few years ago. What she discovered appalled her. Because virtually no studies had controlled for heritability, there was no proof of causation at all in any study. Not even lip service was being paid to this omission: correlation was being routinely presented as causation. Yet in each case, from behaviour genetics studies, there was new, strong evidence against what Rich Harris calls 'the nurture assumption'. Studies of the divorce rate of twins, for example, reveal that genetics accounts for about half of the variation in divorce rate, non-shared environmental factors for another half and shared home environment for nothing at all.[1] In other words, you are no more likely to divorce if reared in a broken home than the average – unless your biological parents divorced. Studies of criminal records of adoptees in Denmark revealed a strong correlation with the criminal record of the biological parent and a very small correlation with the criminal record of the adopting parent – and even that vanished when controlled for peer-group effects, whereby the adopting parents were found to live in more, or less, criminal neighbourhoods according to whether they themselves were criminals.

Indeed, it is now clear that children probably have more non-genetic effect on parents than vice versa. As I argued in the chapter on chromosomes X and Y, it used to be conventional wisdom that distant fathers and over-protective mothers turn sons gay. It is now considered much more likely to be the reverse: perceiving that a son is not fully interested in masculine concerns, the father retreats; the mother compensates by being overprotective. Likewise, it is true that autistic children often have cold mothers; but this is an effect, not a cause: the mother, exhausted and dispirited by years of unrewarding attempts to break through to an autistic child, eventually gives up trying.

Rich Harris has systematically demolished the dogma that has lain, unchallenged, beneath twentieth-century social science: the assumption that parents shape the personality and culture of their children. In Sigmund Freud's psychology, John Watson's behaviourism and

Margaret Mead's anthropology, nurture-determinism by parents was never tested, only assumed. Yet the evidence, from twin studies, from the children of immigrants and from adoption studies, is now staring us in the face: people get their personalities from their genes and from their peers, not from their parents.[1]

In the 1970s, after the publication of E. O. Wilson's book *Sociobiology*, there was a vigorous counter-attack against the idea of genetic influences on behaviour led by Wilson's Harvard colleagues, Richard Lewontin and Stephen Jay Gould. Their favourite slogan, used as a title for one of Lewontin's books, was uncompromisingly dogmatic: 'Not in our genes!' It was at the time still just a plausible hypothesis to assert that genetic influences on behaviour were slight or non-existent. After twenty-five years of studies in behavioural genetics, that view is no longer tenable. Genes do influence behaviour.

Yet even after these discoveries, environment is still massively important – probably in total more important than genes in nearly all behaviours. But a remarkably small part in environmental influence is played by parental influence. This is not to deny that parents matter, or that children could do without them. Indeed, as Rich Harris observes, it is absurd to argue otherwise. Parents shape the home environment and a happy home environment is a good thing in its own right. You do not have to believe that happiness determines personality to agree that it is a good thing to have. But children do not seem to let the home environment influence their personality outside the home, nor to let it influence their personality in later life as an adult. Rich Harris makes the vital observation that we all keep the public and private zones of our lives separate and we do not necessarily take the lessons or the personality from one to the other. We easily 'code-switch' between them. Thus we acquire the language (in the case of immigrants) or accent of our peers, not our parents, for use in the rest of our lives. Culture is transmitted autonomously from each children's peer group to the next and not from parent to child – which is why, for example, the move towards greater adult sexual equality has had zero effect on willing sexual segregation in the playground. As every parent knows, children prefer to imitate peers than parents. Psychology, like sociology and anthropology, has been dominated by those with a strong antipathy to genetic explanations; it can no longer sustain such ignorance.[2]

My point is not to rehearse the nature–nurture debate, which I explored in the chapter on chromosome 6, but to draw attention to the fact that even if the nurture assumption had proved true, it would not have reduced determinism one iota. As it is, by stressing the powerful influence that conformity to a peer group can have on personality, Rich Harris lays bare just how much more alarming social determinism is than genetic. It is brainwashing. Far from leaving room for free will, it rather diminishes it. A child who expresses her own (partly genetic) personality in defiance of her parents' or her siblings' pressures is at least obeying endogenous causality, not somebody else's.

So there is no escape from determinism by appealing to socialisation. Either effects have causes or they do not. If I am timid because of something that happened to me when I was young, that event is no less deterministic than a gene for timidity. The greater mistake is not to equate determinism with genes, but to mistake determinism with inevitability. Said the three authors of *Not in our genes*, Steven Rose, Leon Kamin and Richard Lewontin, 'To the biological determinists the old credo "You can't change human nature" is the alpha and omega of the human condition.' But this equation – determinism equals fatalism – is so well understood to be a fallacy that it is hard to find the straw men that the three critics indict.[3]

The reason the equation of determinism with fatalism is a fallacy is as follows. Suppose you are ill, but you reason that there is no point in calling the doctor because either you will recover, or you won't: in either case, a doctor is superfluous. But this overlooks the possibility that your recovery or lack thereof could be caused by your calling the doctor, or failure to do so. It follows that determinism implies nothing about what you can or cannot do. Determinism looks backwards to the causes of the present state, not forward to the consequences.

Yet the myth persists that genetic determinism is a more implacable kind of fate than social determinism. As James Watson has put it, 'We talk about gene therapy as if it can change someone's fate, but you can also change someone's fate if you pay off their credit card.' The whole point of genetic knowledge is to remedy genetic defects with (mostly non-genetic) interventions. Far from the discoveries of genetic mutations leading to fatalism, I have already cited many examples where they have led to redoubled efforts to ameliorate their effects. As I pointed out in the chapter on chromosome 6, when dyslexia was belatedly recognised as a real, and possibly genetic, condition, the response of parents, teachers and governments was not fatalistic. Nobody said that because it was a genetic condition dyslexia was therefore incurable and from now on children diagnosed with dyslexia would be allowed to remain illiterate. Quite the reverse happened: remedial education for dyslexics was developed, with impressive results. Likewise, as I argued in the chapter on chromosome 11, even psychotherapists have found genetic explanations of shyness helpful in curing it. By reassuring shy people that their shyness is innate and 'real', it somehow helps them overcome it.

Nor does it make sense to argue that biological determinism threatens the case for political freedom. As Sam Brittan has argued, 'the opposite of freedom is coercion, not determinism.'[4] We cherish political freedom because it allows us freedom of personal self-determination, not the other way around. Though we pay lip service to our love of free will, when the chips are down we cling to determinism to save us. In February 1994 an American named Stephen Mobley was convicted of the murder of a pizza-shop manager, John Collins, and sentenced to death. Appealing to have the sentence reduced to life imprisonment, his lawyers offered a genetic defence. Mobley came, they said, from a long pedigree of crooks and criminals. He probably killed Collins because his genes made him do it. 'He' was not responsible; he was a genetically determined automaton.

Mobley was happy to surrender his illusion of free will; he wanted it to be thought that he had none. So does every criminal who uses the defence of insanity or diminished responsibility. So does every jealous spouse who uses the defence of temporary insanity or justifiable rage after murdering an unfaithful partner. So does the unfaithful partner when justifying the infidelity. So does every tycoon who uses the excuse of Alzheimer's disease when accused of fraud against his shareholders. So indeed does a child in the playground who says that his friend made him do it. So does each one of us when we willingly go along with a subtle suggestion from the therapist that we should blame our parents for our present unhappiness. So does a politician who blames social conditions for the crime rate in an area. So does an economist when he asserts that consumers are utility maximisers. So does a biographer when he tries to explain how his subject's character was forged by formative experiences. So does everybody who consults a horoscope. In every case there is a willing, happy and grateful embracing of determinism. Far from loving free will, we seem to be a species that positively leaps to surrender it whenever we can.[5]

Full responsibility for one's actions is a necessary fiction without which the law would flounder, but it is a fiction all the same. To the extent that you act in character you are responsible for your actions; yet acting in character is merely expressing the many determinisms that caused your character. David Hume found himself impaled on this dilemma, subsequently named Hume's fork. Either our actions are determined, in which case we are not responsible for them, or they are random, in which case we are not responsible for them. In either case, common sense is outraged and society impossible to organise.

Christianity has wrestled with these issues for two millennia and theologians of other stripes for much longer. God, almost by definition, seems to deny free will or He would not be omnipotent. Yet Christianity in particular has striven to preserve a concept of free will because, without it, human beings cannot be held accountable for their actions. Without accountability, sin is a mockery and Hell a damnable injustice from a just God. The modern Christian consensus is that

God has implanted free will in us, so that we have a choice of living virtuously or in sin.

Several prominent evolutionary biologists have recently argued that religious belief is an expression of a universal human instinct – that there is in some sense a group of genes for believing in God or gods. (One neuroscientist even claims to have found a dedicated neural module in the temporal lobes of the brain that is bigger or more active in religious believers; hyper-religiosity is a feature of some types of temporal-lobe epilepsy.) A religious instinct may be no more than a by-product of an instinctive superstition to assume that all events, even thunderstorms, have wilful causes. Such a super-stition could have been useful in the Stone Age. When a boulder rolls down the hill and nearly crushes you, it is less dangerous to subscribe to the conspiracy theory that it was pushed by somebody than to assume it was an accident. Our very language is larded with intentionality. I wrote earlier that my genes built me and delegated responsibility to my brain. My genes did nothing of the sort. It all just happened.

E. O. Wilson even argues, in his book *Consilience*,[6] that morality is the codified expression of our instincts, and that what is right is indeed – despite the naturalistic fallacy – derived from what comes naturally. This leads to the paradoxical conclusion that belief in a god, being natural, is therefore correct. Yet Wilson himself was reared a devout Baptist and is now an agnostic, so he has rebelled against a deterministic instinct. Likewise, Steven Pinker, by remaining childless while subscribing to the theory of the selfish gene, has told his selfish genes to 'go jump in a lake'.

So even determinists can escape determinism. We have a paradox. Unless our behaviour is random, then it is determined. If it is determined, then it is not free. And yet we feel, and demonstrably are, free. Charles Darwin described free will as a delusion caused by our inability to analyse our own motives. Modern Darwinists such as Robert Trivers have even argued that deceiving ourselves about such matters is itself an evolved adaptation. Pinker has called free will 'an idealisation of human beings that makes the ethics game playable'. The writer Rita Carter calls it an illusion hard-wired into the mind. The philosopher Tony Ingram calls free will something that we assume other people have – we seem to have an inbuilt bias to ascribe free will to everybody and everything about us, from recalcitrant outboard motors to recalcitrant children equipped with our genes.[7]

I would like to think that we can get a little closer to resolving the paradox than that. Recall that, when discussing chromosome 10, I described how the stress response consists of genes at the whim of the social environment, not vice versa. If genes can affect behaviour and behaviour can affect genes, then the causality is circular. And in a system of circular feedbacks, hugely unpredictable results can follow from simple deterministic processes.

This kind of notion goes under the name of chaos theory. Much as I hate to admit it, the physicists have got there first. Pierre-Simon de LaPlace, the great French mathematician of the eighteenth century, once mused that if, as a good Newtonian, he could know the positions and the motions of every atom in the universe, he could predict the future. Or rather, he suspected that he could not know the future, but he wondered why not. It is fashionable to say that the answer lies at the subatomic level, where we now know that there are quantum-mechanical events that are only statistically predictable and the world is not made of Newtonian billiard balls. But that is not much help because Newtonian physics is actually a pretty good description of events at the scale at which we live and nobody seriously believes that we rely, for our free will, on the probabilistic scaffolding of Heisenberg's uncertainty principle. To put the reason bluntly: in deciding to write this chapter this afternoon, my brain did not play dice. To act randomly is not the same thing as to act freely – in fact, quite the reverse.[8]

Chaos theory provides a better answer to LaPlace. Unlike quantum physics, it does not rest on chance. Chaotic systems, as defined by mathematicians, are determined, not random. But the theory holds that even if you know all the determining factors in a system, you may not be able to predict the course it will take, because of the way different causes can interact with each other. Even simply determined systems can behave chaotically. They do so partly because of reflexivity, whereby one action affects the starting conditions of the next action, so small effects become larger causes. The trajectory of the stock market index, the future of the weather and the 'fractal geometry' of a coastline are all chaotic systems: in each case, the broad outline or course of events is predictable, but the precise details are not. We know it will be colder in winter than summer, but we cannot tell whether it will snow next Christmas Day.

Human behaviour shares these characteristics. Stress can alter the expression of genes, which can affect the response to stress and so on. Human behaviour is therefore unpredictable in the short term, but broadly predictable in the long term. Thus at any instant in the day, I can choose not to consume a meal. I am free not to eat. But over the course of the day it is almost a certainty that I will eat. The timing of my meal may depend on many things – my hunger (partly dictated by my genes), the weather (chaotically determined by myriad external factors), or somebody else's decision to ask me out to lunch (he being a deterministic being over whom I have no control). This interaction of genetic and external influences makes my behaviour unpredictable, but not undetermined. In the gap between those words lies freedom.

We can never escape from determinism, but we can make a distinction between good determinisms and bad ones – free ones and unfree ones. Suppose that I am sitting in the laboratory of Shin Shimojo at the California Institute of Technology and he is at this very moment prodding with an electrode a part of my brain somewhere close to the anterior cingulate sulcus. Since the control of 'voluntary' movement is in this general area, he might be responsible for me making a movement that would, to me, have all the appearance of volition. Asked why I had moved my arm, I would almost certainly reply with conviction that it was a voluntary decision. Professor Shimojo would know better (I hasten to add that this is still a thought experiment suggested to me by Shimojo, not a real one). It was not the fact that my movement was determined that contradicted my illusion of freedom; it was the fact that it was determined from outside by somebody else.

The philosopher A. J. Ayer put it this way:[9]

If I suffered from a compulsive neurosis, so that I got up and walked across the room, whether I wanted to or not, or if I did so because somebody else compelled me, then I should not be acting freely. But if I do it now, I shall be acting freely, just because these conditions do not obtain; and the fact that my action may nevertheless have a cause is, from this point of view, irrelevant.

A psychologist of twins, Lyndon Eaves, has made a similar point:[10]

Freedom is the ability to stand up and transcend the limitations of the environment. That capacity is something that natural selection has placed in us, because it's adaptive . . . If you're going to be pushed around, would you rather be pushed around by your environment, which is not you, or by your genes, which in some sense is who you are.

Freedom lies in expressing your own determinism, not somebody else's. It is not the determinism that makes a difference, but the ownership. If freedom is what we prefer, then it is preferable to be determined by forces that originate in ourselves and not in others. Part of our revulsion at cloning originates in the fear that what is uniquely ours could be shared by another. The single-minded obsession of the genes to do the determining in their own body is our strongest bulwark against loss of freedom to external causes. Do you begin to see why I facetiously flirted with the idea of a gene for free will? A gene for free will would not be such a paradox because it would locate the source of our behaviour inside us, where others cannot get at it. Of course, there is no single gene, but instead there is something infinitely more uplifting and magnificent: a whole human nature, flexibly preordained in our chromosomes and idiosyncratic to each of us. Everybody has a unique and different, endogenous nature. A self.

BIBLIOGRAPHY AND NOTES

The literature of genetics and molecular biology is gargantuan and out of date. As it is published, each book, article or scientific paper requires updating or revising, so fast is new knowledge being minted (the same applies to my book). So many scientists are now working in the field that it is almost impossible even for many of them to keep up with each other's work. When writing this book, I found that frequent trips to the library and conversations with scientists were not enough. The new way to keep abreast was to surf the Net.

The best repository of genetic knowledge is found at Victor McCusick's incomparable website known as OMIM, for Online Mendelian Inheritance in Man. Found at www3.ncbi.nlm.nih.gov:80/htbin_post/Omim, it includes a separate essay with sources on every human gene that has been mapped or sequenced, and it is updated very regularly – an almost overwhelming task. The Weizmann Institute in Israel has another excellent website with 'gene-cards' summarising what is known about each gene and links to other relevant websites: //bioinformatics.weizmann.ac.il/cards.

But these websites give only summaries of knowledge and they are not for the faint-hearted: there is much jargon and assumed knowledge, which will defeat many amateurs. They also concentrate on the relevance of each gene for inherited disorders, thus compounding the problem that I have tried to combat in this book: the impression that the main function of genes is to cause diseases.

I have relied heavily on textbooks, therefore, to supplement and explain the latest knowledge. Some of the best are Tom Strachan and Andrew Read's *Human molecular genetics* (Bios Scientific Publishers, 1996), Robert Weaver and Philip Hedrick's *Basic genetics* (William C. Brown, 1995), David Micklos and Greg Freyer's *DNA science* (Cold Spring Harbor Laboratory Press, 1990) and Benjamin Lewin's *Genes VI* (Oxford University Press, 1997).

As for more popular books about the genome in general, I recommend Christopher Wills's *Exons, introns and talking genes* (Oxford University Press, 1991), Walter Bodmer and Robin McKie's *The book of man* (Little, Brown, 1994) and Steve Jones's *The language of the genes* (Harper Collins, 1993). Also Tom Strachan's *The human genome* (Bios, 1992). All of these are inevitably showing their age, though.

In each chapter of this book, I have usually relied on one or two main sources, plus a variety of individual scientific papers. The notes that follow are intended to direct the interested reader, who wishes to follow up the subjects, to these sources.

CHROMOSOME 1

The idea that the gene and indeed life itself consists of digital information is found in Richard Dawkins's *River out of Eden* (Weidenfeld and Nicolson, 1995) and in Jeremy Campbell's *Grammatical man* (Allen Lane, 1983). An excellent account of the debates that still rage about the origin of life is found in Paul Davies's *The fifth miracle* (Penguin, 1998). For more detailed information on the RNA world, see Gesteland, R. F. and Atkins, J. F. (eds) (1993). *The RNA world.* Cold Spring Harbor Laboratory Press, Cold Spring Harbor, New York.

1. Darwin, E. (1794). *Zoonomia: or the laws of organic life.* Vol. II, p. 244. Third edition (1801). J. Johnson, London.
2. Campbell, J. (1983). *Grammatical man: information, entropy, language and life.* Allen Lane, London.
3. Schrödinger, E. (1967). *What is life? Mind and matter.* Cambridge University Press, Cambridge.
4. Quoted in Judson, H. F. (1979). *The eighth day of creation.* Jonathan Cape, London.

5. Hodges, A. (1997). *Turing.* Phoenix, London.

6. Campbell, J. (1983). *Grammatical man: information, entropy, language and life.* Allen Lane, London.

7. Joyce, G. F. (1989). RNA evolution and the origins of life. *Nature* 338: 217–24; Unrau, P. J. and Bartel, D. P. (1998). RNA-catalysed nucleotide synthesis. *Nature* 395: 260–63.

8. Gesteland, R. F. and Atkins, J. F. (eds) (1993). *The RNA world.* Cold Spring Harbor Laboratory Press, Cold Spring Harbor, New York.

9. Gold, T. (1992). The deep, hot biosphere. *Proceedings of the National Academy of Sciences of the USA* 89: 6045–49; Gold, T. (1997). An unexplored habitat for life in the universe? *American Scientist* 85: 408–11.

10. Woese, C. (1998). The universal ancestor. *Proceedings of the National Academy of Sciences of the USA* 95: 6854–9.

11. Poole, A. M., Jeffares, D.C and Penny, D. (1998). The path from the RNA world. *Journal of Molecular Evolution* 46: 1–17; Jeffares, D. C., Poole, A. M. and Penny, D. (1998). Relics from the RNA world. *Journal of Molecular Evolution* 46: 18–36.

CHROMOSOME   2

The story of human evolution from an ape ancestor has been told and retold many times. Good recent accounts include: N. T. Boaz's *Eco homo* (Basic Books, 1997), Alan Walker and Pat Shipman's *The wisdom of bones* (Phoenix, 1996), Richard Leakey and Roger Lewin's *Origins revisited* (Little, Brown, 1992) and Don Johanson and Blake Edgar's magnificently illustrated *From Lucy to language* (Weidenfeld and Nicolson, 1996).

1. Kottler, M. J. (1974). From 48 to 46: cytological technique, preconception, and the counting of human chromosomes. *Bulletin of the History of Medicine* 48: 465–502.

2. Young, J. Z. (1950). *The life of vertebrates.* Oxford University Press, Oxford.

3. Arnason, U., Gullberg, A. and Janke, A. (1998). Molecular timing of primate divergences as estimated by two non-primate calibration points. *Journal of Molecular Evolution* 47: 718–27.

4. Huxley, T. H. (1863/1901). *Man's place in nature and other anthropological essays,* p. 153. Macmillan, London.

5. Rogers, A. and Jorde, R. B. (1995). Genetic evidence and modern human origins. *Human Biology* 67: 1–36.
6. Boaz, N. T. (1997). *Eco homo*. Basic Books, New York.
7. Walker, A. and Shipman, P. (1996). *The wisdom of bones*. Phoenix, London.
8. Ridley, M. (1996). *The origins of virtue*. Viking, London.

CHROMOSOME 3

There are many accounts of the history of genetics, of which the best is Horace Judson's *The eighth day of creation* (Jonathan Cape, London, 1979; reprinted by Penguin, 1995). A good account of Mendel's life is found in a novel by Simon Mawer: *Mendel's dwarf* (Doubleday, 1997).

1. Bearn, A. G. and Miller, E. D. (1979). Archibald Garrod and the development of the concept of inborn errors of metabolism. *Bulletin of the History of Medicine* 53: 315–28; Childs, B. (1970). Sir Archibald Garrod's conception of chemical individuality: a modern appreciation. *New England Journal of Medicine* 282: 71–7; Garrod, A. (1909). *Inborn errors of metabolism*. Oxford University Press, Oxford.
2. Mendel, G. (1865). Versuche über Pflanzen-Hybriden. *Verhandlungen des naturforschenden Vereines in Brünn* 4: 3–47. English translation published in the *Journal of the Royal Horticultural Society*, Vol. 26 (1901).
3. Quoted in Fisher, R. A. (1930). *The genetical theory of natural selection*. Oxford University Press, Oxford.
4. Bateson, W. (1909). *Mendel's principles of heredity*. Cambridge University Press, Cambridge.
5. Miescher is quoted in Bodmer, W. and McKie, R. (1994). *The book of man*. Little, Brown, London.
6. Dawkins, R. (1995). *River out of Eden*. Weidenfeld and Nicolson, London.
7. Hayes, B. (1998). The invention of the genetic code. *American Scientist* 86: 8–14.
8. Scazzocchio, C. (1997). Alkaptonuria: from humans to moulds and back. *Trends in Genetics* 13: 125–7; Fernandez-Canon, J. M. and Penalva, M. A. (1995). Homogentisate dioxygenase gene cloned in *Aspergillus*. *Proceedings of the National Academy of Sciences of the USA* 92: 9132–6.

CHROMOSOME 4

For those concerned about inherited disorders such as Huntingdon's disease, the writings of Nancy and Alice Wexler, detailed in the notes below, are essential reading. Stephen Thomas's *Genetic risk* (Pelican, 1986) is a very accessible guide.

1. Thomas, S. (1986). *Genetic risk*. Pelican, London.
2. Gusella, J. F., McNeil, S., Persichetti, F., Srinidhi, J., Novelletto, A., Bird, E., Faber, P., Vonsattel, J.-P., Myers, R. H. and MacDonald, M. E. (1996). Huntington's disease. *Cold Spring Harbor Symposia on Quantitative Biology* 61: 615–26.
3. Huntington, G. (1872). On chorea. *Medical and Surgical Reporter* 26: 317–21.
4. Wexler, N. (1992). Clairvoyance and caution: repercussions from the Human Genome Project. In *The code of codes* (ed. D. Kevles and L. Hood), pp. 211–43. Harvard University Press.
5. Huntington's Disease Collaborative Research Group (1993). A novel gene containing a trinucleotide repeat that is expanded and unstable on Huntington's disease chromosomes. *Cell* 72: 971–83.
6. Goldberg, Y. P. *et al.* (1996). Cleavage of huntingtin by apopain, a proapoptotic cysteine protease, is modulated by the polyglutamine tract. *Nature Genetics* 13: 442–9; DiFiglia, M., Sapp, E., Chase, K. O., Davies, S. W., Bates, G. P., Vonsattel, J. P. and Aronin, N. (1997). Aggregation of huntingtin in neuronal intranuclear inclusions and dystrophic neurites in brain. *Science* 277: 1990–93.
7. Kakiuza, A. (1998). Protein precipitation: a common etiology in neurodegenerative disorders? *Trends in genetics* 14: 398–402.
8. Bat, O., Kimmel, M. and Axelrod, D. E. (1997). Computer simulation of expansions of DNA triplet repeats in the fragile-X syndrome and Huntington's disease. *Journal of Theoretical Biology* 188: 53–67.
9. Schweitzer, J. K. and Livingston, D. M. (1997). Destabilisation of CAG trinucleotide repeat tracts by mismatch repair mutations in yeast. *Human Molecular Genetics* 6: 349–55.
10. Mangiarini, L. (1997). Instability of highly expanded CAG repeats in mice transgenic for the Huntington's disease mutation. *Nature Genetics* 15: 197–200; Bates, G. P., Mangiarini, L., Mahal, A. and Davies, S. W. (1997).

Transgenic models of Huntington's disease. *Human Molecular Genetics* 6: 1633–7.

11. Chong, S. S. *et al.* (1997). Contribution of DNA sequence and CAG size to mutation frequencies of intermediate alleles for Huntington's disease: evidence from single sperm analyses. *Human Molecular Genetics* 6: 301–10.

12. Wexler, N. S. (1992). The Tiresias complex: Huntington's disease as a paradigm of testing for late-onset disorders. *FASEB Journal* 6: 2820–25.

13. Wexler, A. (1995). *Mapping fate.* University of California Press, Los Angeles.

CHROMOSOME 5

One of the best books about gene hunting is William Cookson's *The gene hunters: adventures in the genome jungle* (Aurum Press, 1994). Cookson is one of my main sources of information on asthma genes.

1. Hamilton, G. (1998). Let them eat dirt. *New Scientist,* 18 July 1998: 26–31; Rook, G. A. W. and Stanford, J. L. (1998). Give us this day our daily germs. *Immunology Today* 19: 113–16.

2. Cookson, W. (1994). *The gene hunters: adventures in the genome jungle.* Aurum Press, London.

3. Marsh, D. G. *et al.* (1994). Linkage analysis of IL4 and other chromosome 5q31.1 markers and total serum immunoglobulin-E concentrations. *Science* 264: 1152–6.

4. Martinez, F. D. *et al.* (1997). Association between genetic polymorphism of the beta-2-adrenoceptor and response to albuterol in children with or without a history of wheezing. *Journal of Clinical Investigation* 100: 3184–8.

CHROMOSOME 6

The story of Robert Plomin's search for genes that influence intelligence will be told in a forthcoming book by Rosalind Arden. Plomin's textbook on *Behavioral genetics* is an especially readable introduction to the field (third edition, W. H. Freeman, 1997). Stephen Jay Gould's *Mismeasure of man*

(Norton, 1981) is a good account of the early history of eugenics and IQ. Lawrence Wright's *Twins: genes, environment and the mystery of identity* (Weidenfeld and Nicolson, 1997) is a delightful read.

1. Chorney, M. J., Chorney, K., Seese, N., Owen, M. J., Daniels, J., McGuffin, P., Thompson, L. A., Detterman, D. K., Benbow, C., Lubinski, D., Eley, T. and Plomin, R. (1998). A quantitative trait locus associated with cognitive ability in children. *Psychological Science* 9: 1–8.

2. Galton, F. (1883). *Inquiries into human faculty*. Macmillan, London.

3. Goddard, H. H. (1920), quoted in Gould, S. J. (1981). *The mismeasure of man*. Norton, New York.

4. Neisser, U. *et al.* (1996). Intelligence: knowns and unknowns. *American Psychologist* 51: 77–101.

5. Philpott, M. (1996). Genetic determinism. In Tam, H. (ed.), *Punishment, excuses and moral development*. Avebury, Aldershot.

6. Wright, L. (1997). *Twins: genes, environment and the mystery of identity*. Weidenfeld and Nicolson, London.

7. Scarr, S. (1992). Developmental theories for the 1990s: development and individual differences. *Child Development* 63: 1–19.

8. Daniels, M., Devlin, B. and Roeder, K. (1997). Of genes and IQ. In Devlin, B., Fienberg, S. E., Resnick, D. P. and Roeder, K. (eds), *Intelligence, genes and success*. Copernicus, New York.

9. Herrnstein, R. J. and Murray, C. (1994). *The bell curve*. The Free Press, New York.

10. Haier, R. *et al.* (1992). Intelligence and changes in regional cerebral glucose metabolic rate following learning. *Intelligence* 16: 415–26.

11. Gould, S. J. (1981). *The mismeasure of man*. Norton, New York.

12. Furlow, F. B., Armijo-Prewitt, T., Gangestead, S. W. and Thornhill, R. (1997). Fluctuating asymmetry and psychometric intelligence. *Proceedings of the Royal Society of London, Series B* 264: 823–9.

13. Neisser, U. (1997). Rising scores on intelligence tests. *American Scientist* 85: 440–47.

CHROMOSOME   7

Evolutionary psychology, the theme of this chapter, is explored in several books, including Jerome Barkow, Leda Cosmides and John Tooby's *The adapted mind* (Oxford University Press, 1992), Robert Wright's *The moral animal* (Pantheon, 1994), Steven Pinker's *How the mind works* (Penguin, 1998) and my own *The red queen* (Viking, 1993). The origin of human language is explored in Steven Pinker's *The language instinct* (Penguin, 1994) and Terence Deacon's *The symbolic species* (Penguin, 1997).

1. For the death of Freudianism: Wolf, T. (1997). Sorry but your soul just died. *The Independent on Sunday*, 2 February 1997. For the death of Meadism: Freeman, D. (1983). Margaret Mead and Samoa: the making and unmaking of an anthropological myth. Harvard University Press, Cambridge, MA; Freeman, D. (1997). *Frans Boas and 'The flower of heaven'*. Penguin, London. For the death of behaviourism: Harlow, H. F., Harlow, M. K. and Suomi, S. J. (1971). From thought to therapy: lessons from a primate laboratory. *American Scientist* 59: 538–49.

2. Pinker, S. (1994). *The language instinct: the new science of language and mind*. Penguin, London.

3. Dale, P. S., Simonoff, E., Bishop, D. V. M., Eley, T. C., Oliver, B., Price, T. S., Purcell, S., Stevenson, J. and Plomin, R. (1998). Genetic influence on language delay in two-year-old children. *Nature Neuroscience* 1: 324–8; Paulesu, E. and Mehler, J. (1998). Right on in sign language. *Nature* 392: 233–4.

4. Carter, R. (1998). *Mapping the mind*. Weidenfeld and Nicolson, London.

5. Bishop, D. V. M., North, T. and Donlan, C. (1995). Genetic basis of specific language impairment: evidence from a twin study. *Developmental Medicine and Child Neurology* 37: 56–71.

6. Fisher, S. E., Vargha-Khadem, F., Watkins, K. E., Monaco, A. P. and Pembrey, M. E. (1998). Localisation of a gene implicated in a severe speech and language disorder. *Nature Genetics* 18: 168–70.

7. Gopnik, M. (1990). Feature-blind grammar and dysphasia. *Nature* 344: 715.

8. Fletcher, P. (1990). Speech and language deficits. *Nature* 346: 226; Vargha-Khadem, F. and Passingham, R. E. (1990). Speech and language deficits. *Nature* 346: 226.

9. Gopnik, M., Dalakis, J., Fukuda, S. E., Fukuda, S. and Kehayia, E. (1996). Genetic language impairment: unruly grammars. In Runciman, W. G., Maynard Smith, J. and Dunbar, R. I. M. (eds), *Evolution of social behaviour patterns in primates and man*, pp. 223–49. Oxford University Press, Oxford; Gopnik, M. (ed.) (1997). *The inheritance and innateness of grammars*. Oxford University Press, Oxford.

10. Gopnik, M. and Goad, H. (1997). What underlies inflectional error patterns in genetic dysphasia? *Journal of Neurolinguistics* 10: 109–38; Gopnik, M. (1999). Familial language impairment: more English evidence. *Folia Phonetica et Logopaedia* 51: in press. Myrna Gopnik, e-mail correspondence with the author, 1998.

11. Associated Press, 8 May 1997; Pinker, S. (1994). *The language instinct: the new science of language and mind*. Penguin, London.

12. Mineka, S. and Cook, M. (1993). Mechanisms involved in the observational conditioning of fear. *Journal of Experimental Psychology, General* 122: 23–38.

13. Dawkins, R. (1986). *The blind watchmaker*. Longman, Essex.

CHROMOSOMES X AND Y

The best place to find out more about intragenomic conflict is in Michael Majerus, Bill Amos and Gregory Hurst's textbook *Evolution: the four billion year war* (Longman, 1996) and W. D. Hamilton's *Narrow roads of gene land* (W. H. Freeman, 1995). For the studies that led to the conclusion that homosexuality was partly genetic, see Dean Hamer and Peter Copeland's *The science of desire* (Simon and Schuster, 1995) and Chandler Burr's *A separate creation: how biology makes us gay* (Bantam Press, 1996).

1. Amos, W. and Harwood, J. (1998). Factors affecting levels of genetic diversity in natural populations. *Philosophical Transactions of the Royal Society of London, Series B* 353: 177–86.

2. Rice, W. R. and Holland, B. (1997). The enemies within: intergenomic conflict, interlocus contest evolution (ICE), and the intraspecific Red Queen. *Behavioral Ecology and Sociobiology* 41: 1–10.

3. Majerus, M., Amos, W. and Hurst, G. (1996). *Evolution: the four billion year war*. Longman, Essex.

4. Swain, A., Narvaez, V., Burgoyne, P., Camerino, G. and Lovell-Badge, R. (1998). Dax1 antagonises sry action in mammalian sex determination. *Nature* 391: 761–7.

5. Hamilton, W. D. (1967). Extraordinary sex ratios. *Science* 156: 477–88.

6. Amos, W. and Harwood, J. (1998). Factors affecting levels of genetic diversity in natural populations. *Philosophical Transactions of the Royal Society of London, Series B* 353: 177–86.

7. Rice, W. R. (1992). Sexually antagonistic genes: experimental evidence. *Science* 256: 1436–9.

8. Haig, D. (1993). Genetic conflicts in human pregnancy. *Quarterly Review of Biology* 68: 495–531.

9. Holland, B. and Rice, W. R. (1998). Chase-away sexual selection: antagonistic seduction versus resistance. *Evolution* 52: 1–7.

10. Rice, W. R. and Holland, B. (1997). The enemies within: intergenomic conflict, interlocus contest evolution (ICE), and the intraspecific Red Queen. *Behavioral Ecology and Sociobiology* 41: 1–10.

11. Hamer, D. H., Hu, S., Magnuson, V. L., Hu, N. *et al.* (1993). A linkage between DNA markers on the X chromosome and male sexual orientation. *Science* 261: 321–7; Pillard, R. C. and Weinrich, J. D. (1986). Evidence of familial nature of male homosexuality. *Archives of General Psychiatry* 43: 808–12.

12. Bailey, J. M. and Pillard, R. C. (1991). A genetic study of male sexual orientation. *Archives of General Psychiatry* 48: 1089–96; Bailey, J. M. and Pillard, R. C. (1995). Genetics of human sexual orientation. *Annual Review of Sex Research* 6: 126–50.

13. Hamer, D. H., Hu, S., Magnuson, V. L., Hu, N. *et al.* (1993). A linkage between DNA markers on the X chromosome and male sexual orientation. *Science* 261: 321–7.

14. Bailey, J. M., Pillard, R. C., Dawood, K., Miller, M. B., Trivedi, S., Farrer, L. A. and Murphy, R. L.; in press. A family history study of male sexual orientation: no evidence for X-linked transmission. *Behaviour Genetics.*

15. Blanchard, R. (1997). Birth order and sibling sex ratio in homosexual versus heterosexual males and females. *Annual Review of Sex Research* 8: 27–67.

16. Blanchard, R. and Klassen, P. (1997). H-Y antigen and homosexuality in men. *Journal of Theoretical Biology* 185: 373–8; Arthur, B. I., Jallon, J.-M., Caflisch, B., Choffat, Y. and Nothiger, R. (1998). Sexual behaviour in *Drosophila* is irreversibly programmed during a critical period. *Current Biology* 8: 1187–90.

17. Hamilton, W. D. (1995). *Narrow roads of gene land*, Vol. 1. W. H. Freeman, Basingstoke.

CHROMOSOME 8

Again, one of the best sources on mobile genetic elements is the textbook by Michael Majerus, Bill Amos and Gregory Hurst: *Evolution: the four billion year war* (Longman, 1996). A good account of the invention of genetic fingerprinting is in Walter Bodmer and Robin McKie's *The book of man* (Little, Brown, 1994). Sperm competition theory is explored in Tim Birkhead and Anders Moller's *Sperm competition in birds* (Academic Press, 1992).

1. Susan Blackmore explained this trick in her article 'The power of the meme meme' in the *Skeptic*, Vol. 5 no. 2, p. 45.
2. Kazazian, H. H. and Moran, J. V. (1998). The impact of L1 retrotransposons on the human genome. *Nature Genetics* 19: 19–24.
3. Casane, D., Boissinot, S., Chang, B. H. J., Shimmin, L. C. and Li, W. H. (1997). Mutation pattern variation among regions of the primate genome. *Journal of Molecular Evolution* 45: 216–26.
4. Doolittle, W. F. and Sapienza, C. (1980). Selfish genes, the phenotype paradigm and genome evolution. *Nature* 284: 601–3; Orgel, L. E. and Crick, F. H. C. (1980). Selfish DNA: the ultimate parasite. *Nature* 284: 604–7.
5. McClintock, B. (1951). Chromosome organisation and genic expression. *Cold Spring Harbor Symposia on Quantitative Biology* 16: 13–47.
6. Yoder, J. A., Walsh, C. P. and Bestor, T. H. (1997). Cytosine methylation and the ecology of intragenomic parasites. *Trends in Genetics* 13: 335–40; Garrick, D., Fiering, S., Martin, D. I. K. and Whitelaw, E. (1998). Repeat-induced gene silencing in mammals. *Nature Genetics* 18: 56–9.
7. Jeffreys, A. J., Wilson, V. and Thein, S. L. (1985). Hypervariable 'minisatellite' regions in human DNA. *Nature* 314: 67–73.
8. Reilly, P. R. and Page, D. C. (1998). We're off to see the genome. *Nature Genetics* 20: 15–17; *New Scientist*, 28 February 1998, p. 20.
9. See *Daily Telegraph*, 14 July 1998, and *Sunday Times*, 19 July 1998.
10. Ridley, M. (1993). *The Red Queen: sex and the evolution of human nature*. Viking, London.

CHROMOSOME   9

Randy Nesse and George Williams's *Evolution and healing* (Weidenfeld and Nicolson, 1995) is the best introduction to Darwinian medicine and the interplay between genes and pathogens.

1. Crow, J. F. (1993). Felix Bernstein and the first human marker locus. *Genetics* 133: 4–7.
2. Yamomoto, F., Clausen, H., White, T., Marken, S. and Hakomori, S. (1990). Molecular genetic basis of the histo-blood group ABO system. *Nature* 345: 229–33.
3. Dean, A. M. (1998). The molecular anatomy of an ancient adaptive event. *American Scientist* 86: 26–37.
4. Gilbert, S. C., Plebanski, M., Gupta, S., Morris, J., Cox, M., Aidoo, M., Kwiatowski, D., Greenwood, B. M., Whittle, H. C. and Hill, A. V. S. (1998). Association of malaria parasite population structure, HLA and immunological antagonism. *Science* 279: 1173–7; also A. Hill, personal communication.
5. Pier, G. B. *et al.* (1998). *Salmonella typhi* uses CFTR to enter intestinal epithelial cells. *Nature* 393: 79–82.
6. Hill, A. V. S. (1996). Genetics of infectious disease resistance. *Current Opinion in Genetics and Development* 6: 348–53.
7. Ridley, M. (1997). *Disease.* Phoenix, London.
8. Cavalli-Sforza, L. L. and Cavalli-Sforza, F. (1995). *The great human diasporas.* Addison Wesley, Reading, Massachusetts.
9. Wederkind, C. and Füri, S. (1997). Body odour preferences in men and women: do they aim for specific MHC combinations or simple heterogeneity? *Proceedings of the Royal Society of London, Series B* 264: 1471–9.
10. Hamilton, W. D. (1990). Memes of Haldane and Jayakar in a theory of sex. *Journal of Genetics* 69: 17–32.

CHROMOSOME 10

The tricky subject of psychoneuroimmunology is explored by Paul Martin's *The sickening mind* (Harper Collins, 1997).

1. Martin, P. (1997). *The sickening mind: brain, behaviour, immunity and disease.* Harper Collins, London.
2. Becker, J. B., Breedlove, M. S. and Crews, D. (1992). *Behavioral endocrinology.* MIT Press, Cambridge, Massachusetts.
3. Marmot, M. G., Davey Smith, G., Stansfield, S., Patel, C., North, F. and Head, J. (1991). Health inequalities among British civil servants: the Whitehall II study. *Lancet* 337: 1387–93.
4. Sapolsky, R. M. (1997). *The trouble with testosterone and other essays on the biology of the human predicament.* Touchstone Press, New York.
5. Folstad, I. and Karter, A. J. (1992). Parasites, bright males and the immunocompetence handicap. *American Naturalist* 139: 603–22.
6. Zuk, M. (1992). The role of parasites in sexual selection: current evidence and future directions. *Advances in the Study of Behavior* 21: 39–68.

CHROMOSOME 11

Dean Hamer has both done the research and written the books on personality genetics and the search for genetic markers that correlate with personality differences. His book, with Peter Copeland, is *Living with our genes* (Doubleday, 1998).

1. Hamer, D. and Copeland, P. (1998). *Living with our genes.* Doubleday, New York.
2. Efran, J. S., Greene, M. A. and Gordon, D. E. (1998). Lessons of the new genetics. *Family Therapy Networker* 22 (March/April 1998): 26–41.
3. Kagan, J. (1994). *Galen's prophecy: temperament in human nature.* Basic Books, New York.
4. Wurtman, R. J. and Wurtman, J. J. (1994). Carbohydrates and depression. In Masters, R. D. and McGuire, M. T. (eds), *The neurotransmitter revolution,* pp.96–109. Southern Illinois University Press, Carbondale and Edwardsville.

5. Kaplan, J. R., Fontenot, M. B., Manuck, S. B. and Muldoon, M. F. (1996). Influence of dietary lipids on agonistic and affiliative behavior in *Macaca fascicularis. American Journal of Primatology* 38: 333–47.

6. Raleigh, M. J. and McGuire, M. T. (1994). Serotonin, aggression and violence in vervet monkeys. In Masters, R. D. and McGuire, M. T. (eds), *The neurotransmitter revolution*, pp. 129–45. Southern Illinois University Press, Carbondale and Edwardsville.

CHROMOSOME 12

The story of homeotic genes and the way in which they have opened up the study of embryology is told in two recent textbooks: *Principles of development* by Lewis Wolpert (with Rosa Beddington, Jeremy Brockes, Thomas Jessell, Peter Lawrence and Elliot Meyerowitz) (Oxford University Press, 1998), and *Cells, embryos and evolution* by John Gerhart and Marc Kirschner (Blackwell, 1997).

1. Bateson, W. (1894). *Materials for the study of variation.* Macmillan, London.

2. Tautz, D. and Schmid, K. J. (1998). From genes to individuals: developmental genes and the generation of the phenotype. *Philosophical Transactions of the Royal Society of London, Series B* 353: 231–40.

3. Nüsslein-Volhard, C. and Wieschaus, E. (1980). Mutations affecting segment number and polarity in *Drosophila. Nature* 287: 795–801.

4. McGinnis, W., Garber, R. L., Wirz, J., Kuriowa, A. and Gehring, W. J. (1984). A homologous protein coding sequence in *Drosophila* homeotic genes and its conservation in other metazoans. *Cell* 37: 403–8; Scott, M. and Weiner, A. J. (1984). Structural relationships among genes that control development: sequence homology between the *Antennapedia, Ultrabithorax* and *fushi tarazu* loci of *Drosophila. Proceedings of the National Academy of Sciences of the USA* 81: 4115–9.

5. Arendt, D. and Nubler-Jung, K. (1994). Inversion of the dorso-ventral axis? *Nature* 371: 26.

6. Sharman, A. C. and Brand, M. (1998). Evolution and homology of the nervous system: cross-phylum rescues of *otd/Otx* genes. *Trends in Genetics* 14: 211–14.

7. Duboule, D. (1995). Vertebrate hox genes and proliferation – an alternative

pathway to homeosis. *Current Opinion in Genetics and Development* 5: 525–8; Krumlauf, R. (1995). Hox genes in vertebrate development. *Cell* 78: 191–201.
8. Zimmer, C. (1998). *At the water's edge*. Free Press, New York.

CHROMOSOME   13

The geography of genes is explored in Luigi Luca Cavalli-Sforza and Francesco Cavalli-Sforza's *The great human diasporas* (Addison Wesley, 1995); some of the same material is also covered in Jared Diamond's *Guns, germs and steel* (Jonathan Cape, 1997).

1. Cavalli-Sforza, L. (1998). The DNA revolution in population genetics. *Trends in Genetics* 14: 60–65.
2. Intriguingly, the genetic evidence generally points to a far more rapid migration rate for women's genes than men's (comparing maternally inherited mitochondria with paternally inherited Y chromosomes) – perhaps eight times as high. This is partly because in human beings, as in other apes, it is generally females that leave, or are abducted from, their native group when they mate. Jensen, M. (1998). All about Adam. *New Scientist*, 11 July 1998: 35–9.
3. Reported in *HMS Beagle: The Biomednet Magazine* (www.biomednet.com/hmsbeagle), issue 20, November 1997.
4. Holden, C. and Mace, R. (1997). Phylogenetic analysis of the evolution of lactose digestion in adults. *Human Biology* 69: 605–28.

CHROMOSOME   14

Two good books on ageing are Steven Austad's *Why we age* (John Wiley and Sons, 1997) and Tom Kirkwood's *Time of our lives* (Weidenfeld and Nicolson, 1999).

1. Slagboom, P. E., Droog, S. and Boomsma, D. I. (1994). Genetic determination of telomere size in humans: a twin study of three age groups. *American Journal of Human Genetics* 55: 876–82.
2. Lingner, J., Hughes, T. R., Shevchenko, A., Mann, M., Lundblad, V. and

Cech, T. R. (1997). Reverse transcriptase motifs in the catalytic subunit of telomerase. *Science* 276: 561–7.

3. Clark, M. S. and Wall, W. J. (1996). *Chromosomes: the complex code*. Chapman and Hall, London.

4. Harrington, L., McPhail, T., Mar, V., Zhou, W., Oulton, R., Bass, M. B., Aruda, I. and Robinson, M. O. (1997). A mammalian telomerase-associated protein. *Science* 275: 973–7; Saito, T., Matsuda, Y., Suzuki, T., Hayashi, A., Yuan, X., Saito, M., Nakayama, J., Hori, T. and Ishikawa, F. (1997). Comparative gene-mapping of the human and mouse TEP-1 genes, which encode one protein component of telomerases. *Genomics* 46: 46–50.

5. Bodnar, A. G. *et al*. (1998). Extension of life-span by introduction of telomerase into normal human cells. *Science* 279: 349–52.

6. Niida, H., Matsumoto, T., Satoh, H., Shiwa, M., Tokutake, Y., Furuichi, Y. and Shinkai, Y. (1998). Severe growth defect in mouse cells lacking the telomerase RNA component. *Nature Genetics* 19: 203–6.

7. Chang, E. and Harley, C. B. (1995). Telomere length and replicative aging in human vascular tissues. *Proceedings of the National Academy of Sciences of the USA* 92: 11190–94.

8. Austad, S. (1997). *Why we age*. John Wiley, New York.

9. Slagboom, P. E., Droog, S. and Boomsma, D. I. (1994). Genetic determination of telomere size in humans: a twin study of three age groups. *American Journal of Human Genetics* 55: 876–82.

10. Ivanova, R. *et al*. (1998). HLA-DR alleles display sex-dependent effects on survival and discriminate between individual and familial longevity. *Human Molecular Genetics* 7: 187–94.

11. The figure of 7,000 genes is given by George Martin, quoted in Austad, S. (1997). *Why we age*. John Wiley, New York.

12. Feng, J. *et al*. (1995). The RNA component of human telomerase. *Science* 269: 1236–41.

CHROMOSOME   15

Wolf Reik and Azim Surani's *Genomic imprinting* (Oxford University Press, 1997) is a good collection of essays on the topic of imprinting. Many books explore gender differences including my own *The Red Queen* (Viking, 1993).

1. Holm, V. *et al.* (1993). Prader–Willi syndrome: consensus diagnostic criteria. *Pediatrics* 91: 398–401.

2. Angelman, H. (1965). 'Puppet' children. *Developmental Medicine and Child Neurology* 7: 681–8.

3. McGrath, J. and Solter, D. (1984). Completion of mouse embryogenesis requires both the maternal and paternal genomes. *Cell* 37: 179–83; Barton, S. C., Surami, M. A. H. and Norris, M. L. (1984). Role of paternal and maternal genomes in mouse development. *Nature* 311: 374–6.

4. Haig, D. and Westoby, M. (1989). Parent-specific gene expression and the triploid endosperm. *American Naturalist* 134: 147–55.

5. Haig, D. and Graham, C. (1991). Genomic imprinting and the strange case of the insulin-like growth factor II receptor. *Cell* 64: 1045–6.

6. Dawson, W. (1965). Fertility and size inheritance in a Peromyscus species cross. *Evolution* 19: 44–55; Mestel, R. (1998). The genetic battle of the sexes. *Natural History* 107: 44–9.

7. Hurst, L. D. and McVean, G. T. (1997). Growth effects of uniparental disomies and the conflict theory of genomic imprinting. *Trends in Genetics* 13: 436–43; Hurst, L. D. (1997). Evolutionary theories of genomic imprinting. In Reik, W. and Surani, A. (eds), *Genomic imprinting*, pp. 211–37. Oxford University Press, Oxford.

8. Horsthemke, B. (1997). Imprinting in the Prader–Willi/Angelman syndrome region on human chromosome 15. In Reik, W. and Surani, A. (eds), *Genomic imprinting*, pp. 177–90. Oxford University Press, Oxford.

9. Reik, W. and Constancia, M. (1997). Making sense or antisense? *Nature* 389: 669–71.

10. McGrath, J. and Solter, D. (1984). Completion of mouse embryogenesis requires both the maternal and paternal genomes. *Cell* 37: 179–83.

11. Jaenisch, R. (1997). DNA methylation and imprinting: why bother? *Trends in Genetics* 13: 323–9.

12. Cassidy, S. B. (1995). Uniparental disomy and genomic imprinting as causes of human genetic disease. *Environmental and Molecular Mutagenesis* 25, Suppl. 26: 13–20; Kishino, T. and Wagstaff, J. (1998). Genomic organisation of the UBE3A/E6-AP gene and related pseudogenes. *Genomics* 47: 101–7.

13. Jiang, Y., Tsai, T.-F., Bressler, J. and Beaudet, A. L. (1998). Imprinting in Angelman and Prader–Willi syndromes. *Current Opinion in Genetics and Development* 8: 334–42.

14. Allen, N. D., Logan, K., Lally, G., Drage, D. J., Norris, M. and Keverne,

E. B. (1995). Distribution of pathenogenetic cells in the mouse brain and their influence on brain development and behaviour. *Proceedings of the National Academy of Sciences of the USA* 92: 10782–6; Trivers, R. and Burt, A. (in preparation), *Kinship and genomic imprinting.*

15. Vines, G. (1997). Where did you get your brains? *New Scientist*, 3 May 1997: 34–9; Lefebvre, L., Viville, S., Barton, S. C., Ishino, F., Keverne, E. B. and Surani, M. A. (1998). Abnormal maternal behaviour and growth retardation associated with loss of the imprinted gene Mest. *Nature Genetics* 20: 163–9.

16. Pagel, M. (1999). Mother and father in surprise genetic agreement. *Nature* 397: 19–20.

17. Skuse, D. H. *et al.* (1997). Evidence from Turner's syndrome of an imprinted locus affecting cognitive function. *Nature* 387: 705–8.

18. Diamond, M. and Sigmundson, H. K. (1997). Sex assignment at birth: long-term review and clinical implications. *Archives of Pediatric and Adolescent Medicine* 151: 298–304.

CHROMOSOME   16

There are no good popular books on the genetics of learning mechanisms. A good textbook is: M. F. Bear, B. W. Connors and M. A. Paradiso's *Neuroscience: exploring the brain* (Williams and Wilkins, 1996).

1. Baldwin, J. M. (1896). A new factor in evolution. *American Naturalist* 30: 441–51, 536–53.

2. Schacher, S., Castelluci, V. F. and Kandel, E. R. (1988). cAMP evokes long-term facilitation in *Aplysia* neurons that requires new protein synthesis. *Science* 240: 1667–9.

3. Bailey, C. H., Bartsch, D. and Kandel, E. R. (1996). Towards a molecular definition of long-term memory storage. *Proceedings of the National Academy of Sciences of the USA* 93: 12445–52.

4. Tully, T., Preat, T., Boynton, S. C. and Del Vecchio, M. (1994). Genetic dissection of consolidated memory in *Drosophila*. *Cell* 79: 39–47; Dubnau, J. and Tully, T. (1998). Gene discovery in *Drosophila*: new insights for learning and memory. *Annual Review of Neuroscience* 21: 407–44.

5. Silva, A. J., Smith, A. M. and Giese, K. P. (1997). Gene targeting and

the biology of learning and memory. *Annual Review of Genetics* 31: 527–46.
6. Davis, R. L. (1993). Mushroom bodies and *Drosophila* learning. *Neuron* 11: 1–14; Grotewiel, M. S., Beck, C. D. O., Wu, K. H., Zhu, X.-R. and Davis, R. L. (1998). Integrin-mediated short-term memory in *Drosophila*. *Nature* 391: 455–60.
7. Vargha-Khadem, F., Gadian, D. G., Watkins, K. E., Connelly, A., Van-Paesschen, W. and Mishkin, M. (1997). Differential effects of early hippo-campal pathology on episodic and semantic memory. *Science* 277: 376–80.

CHROMOSOME 17

The best recent account of cancer research is Robert Weinberg's *One renegade cell* (Weidenfeld and Nicolson, 1998).

1. Hakem, R. *et al.* (1998). Differential requirement for caspase 9 in apoptotic pathways *in vivo. Cell* 94: 339–52.
2. Ridley, M. (1996). *The origins of virtue.* Viking, London; Raff, M. (1998). Cell suicide for beginners. *Nature* 396: 119–22.
3. Cookson, W. (1994). *The gene hunters: adventures in the genome jungle.* Aurum Press, London.
4. *Sunday Telegraph,* 3 May 1998, p. 25.
5. Weinberg, R. (1998). *One renegade cell.* Weidenfeld and Nicolson, London.
6. Levine, A. J. (1997). P53, the cellular gatekeeper for growth and division. *Cell* 88: 323–31.
7. Lowe, S. W. (1995). Cancer therapy and p53. *Current Opinion in Oncology* 7: 547–53.
8. Hüber, A.-O. and Evan, G. I. (1998). Traps to catch unwary oncogenes. *Trends in Genetics* 14: 364–7.
9. Cook-Deegan, R. (1994). *The gene wars: science, politics and the human genome.* W. W. Norton, New York.
10. Krakauer, D. C. and Payne, R. J. H. (1997). The evolution of virus-induced apoptosis. *Proceedings of the Royal Society of London, Series B* 264: 1757–62.
11. Le Grand, E. K. (1997). An adaptationist view of apoptosis. *Quarterly Review of Biology* 72: 135–47.

CHROMOSOME   18

Geoff Lyon and Peter Gorner's blow-by-blow account of the development of gene therapy, *Altered fates* (Norton, 1996) is a good place to start. *Eat your genes* by Stephen Nottingham (Zed Books, 1998) details the history of plant genetic engineering. Lee Silver's *Remaking Eden* (Weidenfeld and Nicolson, 1997) explores the implications of reproductive technologies and genetic engineering in human beings.

1. Verma, I. M. and Somia, N. (1997). Gene therapy – promises, problems and prospects. *Nature* 389: 239–42.
2. Carter, M. H. (1996). Pioneer Hi-Bred: testing for gene transfers. Harvard Business School Case Study N9–597–055.
3. Capecchi, M. R. (1989). Altering the genome by homologous recombination. *Science* 244: 1288–92.
4. First, N. and Thomson, J. (1998). From cows stem therapies? *Nature Biotechnology* 16: 620–21.

CHROMOSOME   19

The promises and perils of genetic screening have been discussed at great length in many books, articles and reports, but few stand out as essential sources of wisdom. Chandler Burr's *A separate creation: how biology makes us gay* (Bantam Press, 1996) is one.

1. Lyon, J. and Gorner, P. (1996). *Altered fates*. Norton, New York.
2. Eto, M., Watanabe, K. and Makino, I. (1989). Increased frequencies of apolipoprotein E2 and E4 alleles in patients with ischemic heart disease. *Clinical Genetics* 36: 183–8.
3. Lucotte, G., Loirat, F. and Hazout, S. (1997). Patterns of gradient of apolipoprotein E allele *4 frequencies in western Europe. *Human Biology* 69: 253–62.
4. Kamboh, M. I. (1995). Apolipoprotein E polymorphism and susceptibility to Alzheimer's disease. *Human Biology* 67: 195–215; Flannery, T. (1998). *Throwim way leg*. Weidenfeld and Nicolson, London.

5. Cook-Degan, R. (1995). *The gene wars: science, politics and the human genome.* Norton, New York.

6. Kamboh, M. I. (1995). Apolipoprotein E polymorphism and susceptibility to Alzheimer's disease. *Human Biology* 67: 195–215; Corder, E. H. *et al.* (1994). Protective effect of apolipoprotein E type 2 allele for late onset Alzheimer disease. *Nature Genetics* 7: 180–84.

7. Bickeboller, H. *et al.* (1997). Apolipoprotein E and Alzheimer disease: genotypic-specific risks by age and sex. *American Journal of Human Genetics* 60: 439–46; Payami, H. *et al.* (1996). Gender difference in apolipoprotein E-associated risk for familial Alzheimer disease: a possible clue to the higher incidence of Alzheimer disease in women. *American Journal of Human Genetics* 58: 803–11; Tang, M.-X. *et al.* (1996). Relative risk of Alzheimer disease and age-at-onset distributions, based on APOE genotypes among elderly African Americans, Caucasians and Hispanics in New York City. *American Journal of Human Genetics* 58: 574–84.

8. Caldicott, F. *et al.* (1998). *Mental disorders and genetics: the ethical context.* Nuffield Council on Bioethics, London.

9. Bickeboller, H. *et al.* (1997). Apolipoprotein E and Alzheimer disease: genotypic-specific risks by age and sex. *American Journal of Human Genetics* 60: 439–46.

10. Maddox, J. (1998). *What remains to be discovered.* Macmillan, London.

11. Cookson, C. (1998). Markers on the road to avoiding illness. *Financial Times*, 3 March 1998, p. 18; Schmidt, K. (1998). Just for you. *New Scientist*, 14 November 1998, p. 32.

12. Wilkie, T. (1996). The people who want to look inside your genes. *Guardian*, 3 October 1996.

CHROMOSOME 20

The story of prions is exceptionally well told in Rosalind Ridley and Harry Baker's *Fatal protein* (Oxford University Press, 1998). I have also drawn on Richard Rhodes's *Deadly feasts* (Simon and Schuster, 1997) and Robert Klitzman's *The trembling mountain* (Plenum, 1998).

1. Prusiner, S. B. and Scott, M. R. (1997). Genetics of prions. *Annual Review of Genetics* 31: 139–75.

2. Brown, D. R. *et al.* (1997). The cellular prion protein binds copper *in vivo*. *Nature* 390: 684–7.

3. Prusiner, S. B., Scott, M. R., DeArmand, S. J. and Cohen, F. E. (1998). Prion protein biology. *Cell* 93: 337–49.

4. Klein, M. A. *et al.* (1997). A crucial role for B cells in neuroinvasive scrapie. *Nature* 390: 687–90.

5. Ridley, R. M. and Baker H. F. (1998). *Fatal protein*. Oxford University Press, Oxford.

CHROMOSOME 2 1

The most thorough history of the eugenics movement, Dan Kevles's *In the name of eugenics* (Harvard University Press, 1985) concentrates mostly on America. For the European scene, John Carey's *The intellectuals and the masses* (Faber and Faber, 1992) is eye-opening.

1. Hawkins, M. (1997). *Social Darwinism in European and American thought*. Cambridge University Press, Cambridge.

2. Kevles, D. (1985). *In the name of eugenics*. Harvard University Press, Cambridge, Massachusetts.

3. Paul, D. B. and Spencer, H. G. (1995). The hidden science of eugenics. *Nature* 374: 302–5.

4. Carey, J. (1992). *The intellectuals and the masses*. Faber and Faber, London.

5. Anderson, G. (1994). The politics of the mental deficiency act. M.Phil. dissertation, University of Cambridge.

6. *Hansard*, 29 May 1913.

7. Wells, H. G., Huxley, J. S. and Wells, G. P. (1931). *The science of life*. Cassell, London.

8. Kealey, T., personal communication; Lindzen, R. (1996). Science and politics: global warming and eugenics. In Hahn, R. W. (ed.), *Risks, costs and lives saved*, pp. 85–103. Oxford University Press, Oxford.

9. King, D. and Hansen, R. (1999). Experts at work: state autonomy, social learning and eugenic sterilisation in 1930s Britain. *British Journal of Political Science* 29: 77–107.

10. Searle, G. R. (1979). Eugenics and politics in Britain in the 1930s. *Annals of Political Science* 36: 159–69.

11. Kitcher, P. (1996). *The lives to come*. Simon and Schuster, New York.

12. Quoted in an interview in the *Sunday Telegraph*, 8 February 1997.

13. Lynn, R. (1996). *Dysgenics: genetic deterioration in modern populations*. Praeger, Westport, Connecticut.

14. Reported in *HMS Beagle: The Biomednet Magazine* (www.biomednet.com/hmsbeagle), issue 20, November 1997.

15. Morton, N. (1998). Hippocratic or hypocritic: birthpangs of an ethical code. *Nature Genetics* 18: 18; Coghlan, A. (1998). Perfect people's republic. *New Scientist*, 24 October 1998, p. 24.

CHROMOSOME   2 2

The most intelligent book on determinism is Judith Rich Harris's *The nurture assumption* (Bloomsbury, 1998). Steven Rose's *Lifelines* (Penguin, 1998) makes the opposing case. Dorothy Nelkin and Susan Lindee's *The DNA mystique* (Freeman, 1995) is worth a look.

1. Rich Harris, J. (1998). *The nurture assumption*. Bloomsbury, London.

2. Ehrenreich, B. and McIntosh, J. (1997). The new creationism. *Nation*, 9 June 1997.

3. Rose, S., Kamin, L. J. and Lewontin, R. C. (1984). *Not in our genes*. Pantheon, London.

4. Brittan, S. (1998). Essays, moral, political and economic. *Hume Papers on Public Policy*, Vol. 6, no. 4. Edinburgh University Press, Edinburgh.

5. Reznek, L. (1997). *Evil or ill? Justifying the insanity defence*. Routledge, London.

6. Wilson, E. O. (1998). *Consilience*. Little, Brown, New York.

7. Darwin's views on free will are quoted in Wright, R. (1994). *The moral animal*. Pantheon, New York.

8. Silver, B. (1998). *The ascent of science*. Oxford University Press, Oxford.

9. Ayer, A. J. (1954). *Philosophical essays*. Macmillan, London.

10. Lyndon Eaves, quoted in Wright, L. (1997). *Twins: genes, environment and mystery of identity*. Weidenfeld and Nicolson, London.

INDEX